Wales Headquarters
Nature Conservancy Council
Plas Penrhos
Ffordd Penrhos
Bangor LL57 2LQ

nature conservancy council

information and library service

☎ Bangor (0248) 55141

THE VANISHING WILD LIFE OF BRITAIN

By the same author

BRITISH GAME
TOWN FOX, COUNTRY FOX
ENQUIRE WITHIN ABOUT ANIMALS
THE WORLD OF FISHES
THE WORLD OF REPTILES
ETC

The Vanishing Wild Life of Britain

Brian Vesey-FitzGerald

MK

MACGIBBON & KEE

First published 1969 by MacGibbon & Kee Ltd
3 Upper James Street Golden Square London W1

Second impression 1969
Printed in Great Britain by
Western Printing Services Ltd
Bristol

SBN 261 61859 8

Acknowledgments

My GRATEFUL thanks are due to the following for their kind permission to quote copyright material: The British Field Sports Society for a passage from *Predatory Birds of Britain*: Cambridge University Press for a passage from *The British Islands and their Vegetation* by Sir Arthur Tansley: Chatto and Windus Ltd. for a passage from *Prehistoric Britain* by Jacquetta and Christopher Hawkes: William Collins, Sons & Co. Ltd. for two passages from *Pesticides and Pollution* by Kenneth Mellanby, and John Murray Ltd. for two passages from *To-morrow's Country-side* by Garth Christian.

Contents

Preface

A PREFACE is, by its very nature, something of an explanation. It has been said that no book needs a preface: that a good book needs no explanation, that no explanation will improve a bad book. I believe that to be true. And yet, I find it necessary to write a preface. For this is a book that does need some explanation.

In the context of the times, the title (a depressing one) will surely lead the reader to expect a dissertation upon the evils of pesticides. In fact, he will find that pesticides are mentioned only towards the very end. The body of the book is concerned not with pesticides, but with man's relationship to wild life through the centuries.

For this, there are two reasons. Firstly, the subject of pesticides has been very thoroughly explored by many people much better qualified to do so than I am. In 1963 the late Rachel Carson, with her classic *Silent Spring*, alerted the world to the dangers involved in the use of toxic chemicals. No book of modern times has had so tremendous an impact. It has been followed by a vast literature on every conceivable aspect of the subject. I give a selection of this literature – admittedly an arbitrary selection: I have not read it all and do not pretend to understand all that I have read – in the bibliography. I have not the scholarship to attempt to add to it.

Secondly, I do not believe that it is the use, or even the abuse, of pesticides that is vitally important to the subject under discussion. It is but one facet of a very much wider problem.

In a sense the title of this book is misleading: for it surely conveys the impression of something new, a recent feature of our national life. In fact, of course, there is nothing new about it. Wild life has always been vanishing from Britain. It has been doing so from the moment that Man began to colonize the island.

That is perfectly true. But now it is a truth with a difference. And the difference lies in Man himself.

BRIAN VESEY-FITZGERALD

THE VANISHING WILD LIFE OF BRITAIN

Chapter 1

IN THE BEGINNING

MEN – men of the genus *Homo*, but of the species *neanderthalensis*, an aberrant branch of the true human stem – were in Britain long before Britain became an island. They were here perhaps 100,000 years ago, during one of the interglacial periods of the Great Ice Age.

That paragraph requires some qualification. Firstly, as to dates. It must be realized that the earlier these are, the less accurate they are. Dates prior to, say, 10,000 B.C. are no more than rough indications. It would be a very pedantic prehistorian who quibbled over a matter of 50,000 years or so. From about 10,000 B.C. to about 2500 B.C. dates (for Britain) are approximations: no more than fairly accurate. From 2500 B.C. onwards they become more and more accurate, but it cannot be too strongly stressed that the first exact historical date in British history is that of Julius Caesar's first expedition in 55 B.C. and that exact historical dating for Britain as a whole begins with the Roman Conquest of A.D. 43. Secondly, as to 'The Great Ice Age'. This is, in fact, a misnomer. During the Pleistocene Epoch of the Quaternary Period – which is the geological period in which we are living – there were four distinct Ice Ages, each of comparatively short duration in terms of geological time, separated by long periods, covering many thousands of years, during which the climate was warm or even hot. The last of these Ice Ages began some 80,000 years ago and ended (in Britain) about 8,000 years ago, when the last of the ice sheets left northern Scotland. This is a mere nothing in terms of geological time, so it may well be that we are living in the fourth interglacial period of the Pleistocene Epoch – not, as is commonly believed, in the Holocene Epoch – and that there is a fifth Ice Age yet to come. It is a sobering thought.

Neanderthal man, or some close relation (there seem to have been about twelve species or races), lived in a Britain which was also the home of the elephant and the hippopotamus and many other animals which we now associate with tropical or sub-tropical climates. Since the immediate predecessors of man in the evolutionary succession were vegetarians, it is not unreasonable to assume that Neanderthal man was largely a vegetarian, living on fruits and berries, nuts, roots, shoots, and probably some insects. He may have trapped one of the smaller animals occasionally: with other members of his family – the total population of Neanderthal man in Britain probably never exceeded a couple of hundred or so – he may occasionally have driven some large, but ailing, creature over a cliff. But his primitive flint implements indicate clearly that he was not a hunter. Almost certainly he lived in fear of, and took good care to avoid, the larger wild beasts. And when the ice returned, covering vegetation and driving the warmth-loving animals southwards, he went too – and to extinction. It may be said that Neanderthal man had no effect on wild life.

And this is equally true of the first representatives of *Homo sapiens*, the men of the Old Stone Age, who came into Britain (still not an island) on the heels of the retreating ice about 50,000 years ago. Palaeolithic man entered a sub-arctic Britain, a land of steppe and tundra with a scrub of dwarf birch and arctic willow, which supported the mammoth and woolly rhinoceros and great herds of bison, reindeer and small horses. It was no land for vegetarians. Palaeolithic man was a hunter, following the herds for his subsistence. Life for him must have been very hard. His contemporaries in south-western France and in northern and eastern Spain decorated the walls of their caves with animal paintings and engravings which are unsurpassed in their inspired naturalism. In regions where the climate was kind and food plentiful there was no necessity for a hunting society always to be on the move. It became static, centred on caves; sending out hunting parties, keeping the women and children at home. There was time for leisure, time to think, time to develop artistic and spiritual potentialities. No such artistic works have been found in British caves occupied by Palaeolithic man. This does not mean that he was less advanced than his cousins to the south: rather is it an indication of the rigour of the life he led. It must be remembered that Britain was then on the very fringe of the habitable world. All the available evidence points to the fact that British caves were occupied only sporadically and for brief periods by small groups. The population probably never exceeded two thousand even at the height of summer in what is now southern England and in winter it was probably never more than two hundred or so. And this was a nomadic population, always on the move, tied to the wandering herds which were its life. It had no time for leisure: no time for the cultural relaxations made possible by the more hospitable conditions which prevailed farther south. And when conditions changed, they did so in such a way as to destroy the very foundation upon which the society of the Old Stone Age was built.

The change was very gradual, of course. Slowly, imperceptibly, the summers grew longer and warmer: slowly, imperceptibly, the ice-sheets shrank and the winters became milder. No doubt successive generations of Palaeolithic man welcomed the slightly warmer summers, the slightly milder winters. No doubt, as the winters became milder, fewer migrated southwards. But as the climate changed, so inevitably did the plant life change. The scrub of dwarf birch and arctic willow gave way to pine: forest began to spread over steppe and tundra, bringing with it new animals. Mammoth and woolly rhinoceros vanished from the wild life of Britain. The great herds of bison and reindeer and small horses were replaced by forest animals: by red deer and elk, by aurochs and wild pig. The economic basis of the nomadic hunting society was totally destroyed. Palaeolithic man gave way to Mesolithic man: the plains man to the forest man.

Mesolithic man probably differed little, if at all, from Palaeolithic man in appearance: it was his mode of life and his tools that differed. This does not mean that he was more civilized: indeed, in many ways he was less advanced. There is, for example, no art of the Middle Stone Age to compare with the cave art of the Old Stone Age. Mesolithic man, like Palaeolithic man before him, was a savage: living off the land. But he was not, as Palaeolithic man had been, a strictly nomadic savage. He settled along the sea coasts, along river valleys, by marshes and meres, and on low, lightly wooded plains near water. He used heavy wood-cutting axes with which he was able to fashion dug-out canoes, flint arrow-heads, and bone implements, easily recognizable as harpoons and fish-spears. Along the coasts he lived in caves and rock shelters, but elsewhere it seems probable that more or less permanent artificial dwellings were often constructed. One such small settlement has been found at Farnham in Surrey near the source of a spring. These dwellings, the oldest man-made dwellings in Britain, were no more than shallow hollows scooped in the ground and roofed with boughs and turves: just one step up from the cave. They were probably not used in summer, but formed the permanent winter settlement of one small community.

Mesolithic man hunted deer, wild pig and small fur-bearing animals in the woods, trapped ducks and geese and many other birds and collected their eggs, and fished with harpoons, barbless bone hooks and, perhaps, with primitive nets. The women collected vegetable foods – hazel nuts were particularly appreciated – and great quantities of shell-fish. All this is evident from the middens. It is also evident that such activities – the population of Mesolithic Britain probably never exceeded three thousand, living in small widely scattered communities – could have had no effect on wild life as a whole.

But towards the end of the Mesolithic period there were further great climatic changes, accompanied by, and perhaps partly caused by, great changes in sea-level. The subsidence of the land, beginning with the encroachment of the North Sea, is thought to have commenced about 7000 B.C. It continued for the next five thousand years or so, until about 2000 B.C., by which date Britain had become an island. Land connection with the Continent is thought to have been severed about 5000 B.C., but the gap separating Britain and France at what is now the Straits of Dover would have been much narrower than it is today for a further couple of thousand years or more.

With these changes in sea-level the climate became much wetter and much warmer, the prevailing winds coming from the south-west and bringing with them rain and mist from the Atlantic. This naturally had a profound effect on the vegetation. The dominant pine forests gave way, on all the heavy soils at least, to forests of oak and alder and many of the river valleys became vast swamps. Since climate controls vegetation and vegetation animal life, there was an equally marked effect on animal life.

The effect of these climatic changes would not, of course, have been so profound had Britain maintained a land bridge, no matter how narrow, with the Continent. In that event natural replacement of forms of animal life wholly unsuited to the new conditions by forms suited to them would have been possible and would have occurred, just as it did at the end of the last Ice Age. The loss of a land link with the Continent meant that those animals unable to adapt themselves to the changed climatic conditions had no means of escape and were, inevitably, doomed to extinction: meant, too, that natural restocking by animals suited to the new conditions was impossible. It was the end of an epoch.

It was the end of an epoch in another sense as well. It was the end of savage Britain. Up to this moment in time man in Britain had been completely subservient to Nature. He had learnt to make primitive tools, it is true, and, learning by bitter experience over thousands of years, to use them with more and more success, but his existence still depended entirely upon his skill in hunting and fishing and upon the skill of his womenfolk in finding the roots and berries necessary to eke out an always uncertain food supply. He had, from the very nature of his life, always to be on the move. Constant movement and the total absence of husbandry prevented any notable increase in population and effectively prevented any change in the way of life. Despite his tools, he lived very much as did the beasts, obeying the dictates of Nature. Now, with the change in climate came a change in men. Mesolithic man gave way to Neolithic man. The men of the New Stone Age were the first civilized men in Britain.

Jacquetta and Christopher Hawkes have pointed out that it was changes in climate that were really responsible for the birth of civilization. I quote, with permission, from their *Prehistoric Britain*:

> Centres existed in the Middle East where animals and plants suitable for domestication by man flourished in conditions that

prompted him to attempt it. The shifting of rain-belts across the world at the end of the Ice Age meant that while in western Europe the open grazing-grounds became wet and tree-ridden, in these more southerly zones they became, little by little, desiccated and desert. This withering of their food supply forced the animals to herd hungrily round the scattered oases, and there they would come into contact with human settlement. Thus it was natural enough that man should have the idea of exercising some control over the habits and movements of the herds that he daily saw about him. At first, no doubt, his authority was slight and his flocks lived in as nearly a wild state as the reindeer of the modern Lapp, but as the animals themselves grew more docile it increased, until with the practice of careful selective breeding man had gone far in his interference with nature.

If it was the hunting male who is likely to have been responsible for the domestication of animals, woman should probably be credited with the great parallel advance of this time – the cultivation of plants. It had always fallen to her to gather the wild vegetable foods, and when she had watched some forgotten handful of seeds sprout into new life, the notion of deliberate sowing might readily suggest itself. Here again selective propagation would in time improve the wild strains, and, granted sufficient rainfall, greatly increase their yield.

Man was no longer subservient to Nature. Now, in some measure, he could assert his will; could make her work for him.

All this, of course, happened long long before Neolithic man first reached Britain. It is now more or less generally accepted that he first settled in southern Britain at some time before 2000 B.C.: perhaps as early as 2500 B.C. By that time Egypt had already entered upon her Middle Empire, had already at least one thousand years of literate urbanized civilization behind her, was on the threshold of one of the greatest periods of artistic achievement – in art, in sculpture, in architecture – in the history of the world. And there were other urban civilizations, notably that of Crete, already well established in the Middle East. These civilizations needed ever-increasing supplies of raw materials and so sent out expeditions to trade with less civilized societies. They would establish trading posts where good sources of supply were found and, inevitably, this meant the introduction of armed men to protect them and to impose their will. As the civilized advanced, so the less civilized retreated. But not empty-handed. They knew nothing of the use of metals – even though it was primarily for metals such as gold, copper and tin that the Egyptians and others established their trading posts – but the first men of the New Stone Age to reach Britain from France (and ultimately from the Mediterranean, for they were of the long-headed Mediterranean stock) had acquired knowledge of husbandry and of the arts that spring from that knowledge. Retreating before a civilization they did not understand, they brought with them civilization.

Chapter 2

THE FIRST FARMERS

THE FIRST Neolithic colonists found Britain heavily wooded, much of it excessively damp and swampy, inhabited by small communities of primitive savages; food-gatherers and hunters. They can have met with little opposition from the native Mesolithic population, whose numbers were too small and whose settlements, on the fringes of the swamps and along the coasts, were too widely scattered to offer any serious resistance. Indeed, it may be doubted if there was any resistance at all. The economies of the two societies, Mesolithic and Neolithic, were so completely different that there can have been but little contact between them. This was not a matter of invasion and conquest, but rather one of peaceful settlement and gradual absorption.

Neolithic man, so far as his tools were concerned, was really no more advanced than Mesolithic man. He, too, used flint and bone implements; and his were not noticeably more sophisticated. The essential difference was that the men of the New Stone Age were farmers, pastoralists and agriculturists. It was they who first introduced domestic animals to Britain, bringing with them cattle, sheep, pigs and the dog. But they did more than that. Mixed farming, pastoralism and agriculture, livestock and arable, demands two things: suitable country and a settled way of life. The settled way of life leads to the establishment of village communities and to an organized society, which in turn gives scope for leisure, the establishment of domestic crafts and the development of spiritual aspirations. Neolithic man brought the village and village crafts to Britain along with his domestic animals.

He settled on the uplands, where the soil was light and pervious. The first settlements would seem to have been on Windmill Hill near Avebury on the Marlborough Downs and at Robin

Hood's Ball near Stonehenge: not, as one might expect from a glance at a map, along the South Downs. In fact, for landsmen in small canoes laden with domestic animals – and a cow in a canoe must pose some problems – a landing on the marsh-fringed (which is what it then was) south coast would have been difficult and hazardous. Much easier to find a river – rivers were the highways of the time – and follow it to its source. The canoes would have come up the Avon, at that time probably still a tributary of the great Solent River, or possibly up the Thames and its tributary, the Kennet. There can, at any rate, be no doubt that the Wiltshire chalk was settled simply because access was easy. The choice, if fortuitous (as it must have been), was fortunate. The Wiltshire Downland, the widest expanse of chalk in Britain, forms the hub, so to speak, of a huge wheel. From it radiate all the principal chalk and limestone ridges of lowland Britain: the Dorset and Hampshire Downs, the North and South Downs, the Mendips, the Chilterns (affording access to the uplands of East Anglia) and the Cotswolds (affording access to the uplands of Northamptonshire and so to the Lincolnshire and Yorkshire Wolds). Along these ridges successive waves of Neolithic colonizers spread in search of fresh grazing and cultivatable land.

Neolithic man was the first to leave his mark upon the face of Britain. His causewayed camps, his flint-mines and stone-axe factories, his Long Barrows and Megalithic tombs, are evidence of a settled, well-organized and vigorous society, with powerful chieftains and revered (or feared) high priests: evidence, too, that the times were peaceful. It would have taken much time and much well-organized labour to build a barrow 200–300 feet long and 50 feet or more wide. It would have taken even longer, made even greater demands upon labour, to build one of the great stone burial chambers with massive megaliths, such as Hetty Pegler's Tump. There can, surely, be no doubt that these huge earthen and stone burial chambers could only have been built in time of peace.

At first sight, the causewayed camps would seem to contradict this. These massive structures of more or less concentric rings of ditches and banks, always situated on hill-tops and commanding extensive views, do look remarkably like military fortifications.

And it is true that some – Maiden Castle in Dorset, for example – were taken over and fortified by later invaders. But the original Neolithic earthwork was not a fortification. The plan of a causewayed camp is always the same: single or multiple ditches with solid causeways of undisturbed earth separating them and dividing them into comparatively short segments. The earth quarried (with antler-picks) from the ditches was thrown up to form banks along their inner margins. In some of the earthworks with multiple ditches there is often a very considerable space between ditches, so that sometimes the innermost embanked enclosure occupies less than a quarter of the total space enclosed by the outermost ditch. The ditches were sometimes eight or more feet deep and the banks were sometimes crowned with stockades. This all sounds like pretty formidable fortification. But always there are those solid causeways of undisturbed earth. These formed gateways through the outer bank, pathways through the inner ones. One causeway might have been defensible, but there is never just one: half a dozen or more are usual. The causewayed camps were livestock corrals: the causeways being left in order to make it easier for the herdsmen to drive their beasts into the embanked enclosures.

There is no evidence that these camps were permanently occupied. It seems probable that the owners of each camp lived in small village settlements in the near neighbourhood of their earthwork. These settlements – composed of beehive huts or of tents made of animal hides; shelters which could be destroyed or taken down at the onset of winter – would have been summer dwelling-places only. Around them the women would cultivate small plots with bone hoes and would grow wheat and flax – the flax probably for its oily seeds rather than for its fibres – while the men tended the animals. This type of agriculture, amounting to no more than scratching of the ground, does not maintain fertility. Fresh ground would have to be broken every two years or so. In the autumn the animals would be herded into the earthwork. Roots were not cultivated, so the surplus stock had to be slaughtered. (This autumn slaughter, by the way, was commonplace in Britain until the beginning of the nineteenth century.) The animals saved for stock would be kept in the embanked enclosures, safe from predatory beasts: their owners would spend the winter in the shelter of the ditches. There is any amount of evidence for the seasonal occupation of these ditches: fragments of pottery, temporary hearths, flint and bone implements, vegetable refuse, large quantities of animal bones, often split for their marrow. Many of these bones are those of domestic animals: cattle, sheep, pigs, goats and dogs. The bones of cattle, and especially those of calves, are very numerous. There was great autumn feasting in the causewayed camps.

The cattle, probably *Bos brachyceros* of southern Europe, were small with moderately long horns. The sheep was almost certainly *Ovis familiaris palustris*, which is thought to have been brought into southern Europe from north Africa. The goat, originally a mountain animal, must have been *Capra prisca*, which is known to have been domesticated early in the Swiss Neolithic period. The pig seems also to have been the same as the pig, *Sus scrofa palustris*, of the Swiss Neolithic period. The dog – complete skeletons have been found at Windmill Hill and at Maiden Castle – was about the same size as a fox terrier, but in appearance must have resembled the dingo. It would have been used as a guard, warning of the approach of humans and of wild beasts by barking, and perhaps also as a herder.

Strangely enough, there is no evidence whatsoever for the domestication of the horse by early Neolithic man in Britain. Horse bones are conspicuous by their absence. They have not been found in the ditches of any of the thirteen known causewayed camps. It is evident that the first Neolithic colonists did not bring the horse with them. Yet it is surely strange that no bones of horses should have been found in these ditches, for their absence suggests that the horse was not then present even as a wild animal. But we know that great herds of small horses were present in Palaeolithic Britain. Are we to assume that those great herds disappeared along with the bison and the mammoth and the woolly rhinoceros? But if the reindeer survived (as it did), there would seem to be no reason why the horse should not have done so. We have here one of the many mysteries of prehistory. What is certain is that the horse was not domesticated by Neolithic man in Britain. The word picture, given in many of the books, of Neolithic man, mounted on a small horse – what we would now call a Celtic-type pony – and accompanied by hunting dogs, pursuing the wild beasts of the forest is colourful, but inaccurate.

This is not to say that Neolithic man did not hunt the wild beasts of the forest. Of course he did: and he would have done so particularly in winter when food was scarce and when there was little work to be done. Palaeolithic and Mesolithic man had no effect upon the numbers of wild beasts because they lived like the beasts, having no thought for the morrow. Neolithic man, the farmer, did not live like the beasts, had thought for the morrow. He was not only the first to introduce domestic animals, the first to establish some sort of village life: he was also the first to affect wild life in Britain.

Let us go back to the beginning again. Why did Neolithic man establish himself on the chalk and the limestone? Today the chalk downs and the limestone hills are almost always bare of trees, forming wide expanses of short, springy turf, perfect pasturage for sheep. It is natural for us to believe that they have always been so. But the ecological botanists tell us that little, if any, of the chalk and limestone uplands was originally free of forest. Beech (long thought to be an introduction, but now known to be native to the chalk of southern Britain), ash, hawthorn, sloe and spindle – and in some places box, juniper and yew – clothed the chalk, the beech being mainly confined to the steep slopes: ash, hawthorn, sloe and spindle to the limestone. The open character which we have come to consider as natural to the chalk and the limestone is, therefore, artificial: the result of man's activity.

The heavy soils of the lowlands at that time – remember, there was then no drainage – were very retentive of water. They supported dense forests of oak with a thick undergrowth of thorn and bramble and holly. Indeed, great areas of lowland – the Weald, for example, and much of the Midland Plain – were virtually impenetrable; huge waterlogged swamps of oakwood. Chalk and limestone soils, on the other hand, are light and pervious. They dry out very quickly, are virtually always dry underfoot. It was for this reason that they – and later the sandy and gravelly soils with their light covering of heath and gorse, birch and rowan – were selected by Neolithic man. They were comparatively easy to clear. It would have taken a very long time and demanded immense labour to clear even an acre of lowland oak forest with stone axes. And such forest could not be destroyed by fire: it was altogether too wet, summer and winter. But the tree covering on the chalk and limestone uplands, on the sandy and gravelly heaths, could be burnt off. It is a mistake to think of Neolithic man chopping away the woods with stone axes. He burnt them.

Now, you cannot clear large areas of woodland and scrub by fire and thereafter keep them clear by grazing and cultivation without having some effect, and surely a very considerable effect, on the wild life and particularly, one would imagine, on birds and insects. Except that we may be sure that Neolithic man was a generous host to fleas and lice, we really know nothing about the insect population of Britain at that time. We know nothing of the butterflies and the moths and the spiders and so on. It may well be that there were then in Britain some species of whose very existence, as species anywhere, we are unaware (insects do not, on the whole, fossilize well) and others of whose existence we are aware, but of whose existence in Britain we are wholly ignorant. It may be that the first clearing activities of Neolithic man hastened some insect species along the road to extinction – who can tell? – but it is much more probable that his pastoral and agricultural activities were beneficial to the insect populations of those areas in which he settled. And, indeed, he may well have brought in species new to Britain with his domestic animals. So far as birds are concerned, it may be doubted whether Neolithic man hastened the extinction of any species: on the contrary, we may be sure that his activities were beneficial to some species. As E. M. Nicholson has pointed out, in his *Birds and Men*, 'sharp-eyed opportunists such as jackdaws and other crows must have been among the first species to adapt themselves to man as a herdsman and later as a cultivator of the more open and less unfriendly tracts of the uplands and elsewhere'. We know, from the meticulous research of James Fisher into fossil records, that the jay and the jackdaw, the crow and the rook, the raven and the magpie were present in Neolithic Britain and we may be quite sure that they would have taken full advantage of man's activities. But we do not, of course, know all the species then present in Britain and we do not know that some which were certainly present in earlier times were present at this time. For example, the skylark is known as an Ice Age fossil: but from that time until the eighth century A.D. there is no record of it. This does not, of course, mean that there were no skylarks in Britain through all those centuries; merely that there is no record of them. If they were present when Neolithic man arrived, then they would have benefited greatly from his clearing activities. If they were not then present (which is possible, for heavily forested country would not have been altogether suitable for them), there can, I think, be little doubt that they would soon have arrived to take advantage of the more open conditions. I think that there can be little doubt that the arrival of Neolithic man would have led to an increase in the numbers of some species and almost certainly to the arrival of others.

But if we know nothing of the insects and not enough of the birds of Neolithic Britain, we do know a good deal about the mammals. Of the carnivores (considering only the larger species) we know that the brown bear, the wolf, the lynx, the wild cat, the fox, the badger and the otter were present. I think it improbable that Neolithic man would have paid much, if any, attention to the otter, but he would, obviously, have killed the others whenever opportunity offered; in self-defence or to protect his domestic animals. He would have paid particular attention to the bear, the wolf and the lynx. But he did not bring about the extinction of any of these.

So much for the carnivores. The ungulates present at the time of the arrival of the first Neolithic colonists were the reindeer, the elk, the red deer, the aurochs, the wild boar, and the giant Irish deer (the Irish 'Elk').

The reindeer would not have concerned him. It had certainly disappeared from southern England before his arrival. But it survived in the northern highlands of Scotland until at least the middle of the twelfth century. The Orkneyinga Saga mentions the hunting of the reindeer by Rognvald and Harald of Orkney, and this is generally agreed to have been about the middle of the twelfth century. But reindeer were certainly extant in Scotland much later than that and may well have survived until the early years of the thirteenth century.

The others would have been hunted for food: though there would have been more than a small element of self-defence in the case of the aurochs, the giant wild ox of legendary ferocity which stood six feet high or more at the shoulder. Neolithic man did not bring about the extinction of any of these and probably had not the slightest effect on their numbers.

But the 'Irish Elk' is another matter. This was really a giant fallow deer, whose antlers weighed up to one hundred pounds and had a span of ten feet or more. It is widely believed that its extinction must have been brought about by the excessive development of the antlers. It is also commonly believed that the dinosaurs became extinct because they developed excessive size. But the dinosaurs, some of them of enormous size, dominated the earth for millions of years, were highly successful creatures, and became extinct for some reason, certainly not related to size, which is not properly understood. The 'Irish Elk' also appears to have been a successful animal over a great part of western Europe and certainly did not become extinct because of the size of its antlers. Change of climate and a consequent change in the nature of its habitat may have had something to do with it. But there seems to be little doubt that it was present, though probably in very small numbers, in southern England when the Neolithic colonists arrived. So large and relatively harmless a beast, so large and splendid a meal, could not possibly have been overlooked. If Neolithic man did not actually kill the last 'Irish Elk' – and who can say? – I think that there can be very little doubt that he hastened it along the road to extinction.

Chapter 3

BRONZE AND IRON

NEOLITHIC man was succeeded by Bronze Age man. This does not mean that Neolithic man abruptly disappeared. The terms Stone Age, Bronze Age, Iron Age are simple and graphic, and valuable for delineating broad periods. But they must be used with discretion. There is a natural tendency to believe that as one age succeeded another, the former just disappeared. This was never the case. Because a new race of men brought metal implements into Britain it does not mean that the use of stone implements suddenly ceased, still less does it mean that the men who used them were wiped out. The implements of one age invariably linger long into the succeeding age: even long into later, thousands of years later, succeeding ages. There are still peoples in the world using bone harpoons and fish-spears not so very different from those used by Mesolithic man in Britain: there are still peoples using the bow and arrow as did the first Bronze Age invaders of Britain: swords of a type evolved in central Europe about 600 B.C. were in use by native tribes in Africa at the end of the nineteenth century: and so on. Long after the arrival of Bronze Age man flint-mines were still being worked in Britain.

At the time of the Bronze Age invasion (about 1900 B.C.) the population of Neolithic Britain probably did not exceed 20,000 and had lived at peace for some 3,000 years. It was wholly unfitted to offer any serious resistance to a well-organized, war-like people of completely different racial stock, armed with more modern weapons. The invaders could have had but little difficulty, no more than a skirmish or two, in overcoming the simple peasant farmers they found occupying the open lands they coveted. They dispossessed them of their pastures, of their herds, and no doubt, since this is the way of conquerors, of the best-looking of their women. But they did not exterminate the men: they put them to work. The Neolithic population became an enslaved population.

The conquest ushered in another long period of peace. It was Bronze Age men who built Avebury and Silbury Hill and Stonehenge. These vast projects could only have been accomplished in a time of prolonged, wholly undisturbed peace: and then only under an exceptionally efficient central government directed by an unquestioned authority. Consider only Stonehenge. Stonehenge would be a considerable engineering achievement in any age, including our own. With the primitive tools available to Bronze Age man it took a very long time to complete and it was not the sort of work that could have been undertaken at all had there been any chance of armed conflict in the land. That there was none is proved by the blue stones of the inner circle. These come from the Prescelly Mountains in Pembrokeshire. Each had to be dug out and then dragged down to the sea. Each had then to be loaded on to a boat or, more probably, slung between two boats or two canoes, and then sailed or paddled all the way round to the mouth of the Avon: and that is a very long way. Each stone had then to be quanted up the river, and then dragged overland again and set up. It was a gigantic undertaking. It could only have been undertaken in times of settled peace. And it could only have been done under the authority of a great king, who had at his command a most efficient civil service. The Bronze Age was, indeed, a golden age.

As one would expect with so well-organized, so firmly established, a society, there was considerable communication and trade with the Continent. Amber was brought from the Baltic, gold and bronze from Ireland, faience beads from Egypt, ornaments from central Europe: a gold cup from Greece and a gold-bound amber disc from Crete have been found in Bronze Age burials. The trade would obviously not have been confined to luxury articles such as these: they are but evidence of a prosperous society. There can be no doubt that domestic animals were also imported: stocks would have had to be strengthened to meet the demands of an increased population and there was, too, a greater area of grazing land available, made possible by a climatic change. It is abundantly evident that shortly before the beginning of the

Bronze Age there was, commencing about 2000 B.C., a change in the climate. The mild moist climate of the 'Atlantic' phase, through which both Mesolithic and Neolithic man had lived, gave way to what is known as the 'sub-Boreal' phase, which continued until almost the end of the Bronze Age, a period of roughly 1,300 years. The climate gradually became drier, with long warm summers. Pines spread over much of the eastern Fenlands and some of the moorland bogs dried up, so that there was a gradual extension of forest to higher levels. The chalk and, though to a lesser extent, the limestone uplands gradually became too dry for cultivation and for that purpose were deserted by Bronze Age man. His burial places (tumuli) are plentiful on the Downs, but most of his actual settlements seem to have been on the lower ground. This means, of course, that he cleared considerable areas of lower ground, mostly along the more accessible and lighter valley lands, and, under the drier conditions prevailing, he would often have used fire to do so. This does not, of course, mean that he deserted the chalk downlands for grazing purposes; so there was much more land available, both on the uplands and on the lower ground, and one can be quite sure that so well-organized a society would have made full use of it.

Though there is no actual evidence of this, additional cattle and pigs must surely have been imported. Probably more goats were imported as well. (The goat is an astonishingly difficult animal to contain and there can be little doubt that there was already a population of feral goats, escapees from the villages of Neolithic man, in existence. Indeed, it may well be that some of the herds of wild goats to be found today in central and north Wales, in the Cheviot Hills, and in the Scottish Highlands have a very ancient ancestry.) We may not be sure about the goats, but we can be quite certain about the sheep. Bronze Age man imported sheep in large numbers, Neolithic man did not know the art of weaving: his women were experts in working leather. The women of the Bronze Age were highly skilled weavers, both in flax and wool. Some of the woollen garments were of extraordinary fineness and there can be no doubt that both the men and the women were fashion-conscious. The men, we know, wore round woollen caps, knee-length kirtles secured by woollen girdles, and long full cloaks which hung from the shoulders. The women also wore cloaks and beneath them short-sleeved jackets and full knee-length skirts. There seems also to have been a more formal dress which reached to the ankles. Now, a lot of sheep would be needed to provide the wool to make such clothes and Neolithic man, we know, had few sheep. It is evident that Bronze Age man must have imported sheep and built up large flocks.

He would have needed to guard them. So it would seem that he must also have imported dogs. Strangely enough, though we have evidence of a Neolithic dog and plenty of evidence for dogs in the succeeding Iron Age, there is no firm evidence for a Bronze Age dog in Britain. But there is any amount of evidence for Bronze Age dogs of different types on the Continent and we know that at this time, and indeed for centuries before this time, dogs had been used by the Egyptians, the Greeks, and the Romans. It is inconceivable that Bronze Age man in Britain did not import dogs.

And he also brought in the horse. This would, of course, have been a small, Celtic-type pony. It would, I think (but this is very much a matter of opinion and one on which there is much argument), have been of a different ancestral stock from the ponies of the Highlands, and may well have resembled the modern Exmoor in size and general appearance. It certainly did not, as I have seen suggested, resemble the Shetland pony. The origin of that breed is a mystery. There is no evidence to suggest that it was in the islands in prehistoric times, nor is there any record of it having been brought to the islands in historic times. And this is strange, for the Shetland pony is not found anywhere else in the world: only in Shetland and Orkney.

Bronze Age man, with his well-organized society, his crop-growing and his large herds of cattle and great flocks of sheep, must have had a very considerable effect on the wild life of the time; at least, on the chalk and limestone uplands and on the valley lands which he cleared and settled. (He did not attempt to clear the damp oakwoods on water-holding clayey soils and probably rarely entered them.) No doubt he made life as uncomfortable as possible for wolves and bears, but his presence and his activities must have been most beneficial to many other forms of wild life. It is not improbable that it was he who introduced the house mouse to Britain – it was not present in Neolithic times – and he must have made life very much easier for a number of seed-eating birds and for such birds as the swallow. The sand martin has never been reliant upon man: it nests in burrows, which it makes for itself. The house martin was originally a cliff-nester (and still is) and has only taken to the outer walls of houses under the eaves in comparatively recent times. But where did the swallows nest before man provided them with sites? It is true that swallows will occasionally nest in the roofs of caves: it is true that they will, very rarely, nest on the branches of trees. But they are, by habit, birds of open country and, particularly, of open country near water. If there were swallows present in Neolithic Britain – and there is no evidence one way or the other – then their numbers must have been very small, simply because there was very little suitable country. Bronze Age man, clearing some of the valley lands and forming settlements of thatched huts, must have given the swallow immense encouragement.

Towards the end of the Bronze Age, roughly about 700 B.C., there was yet another change of climate. The sub-Boreal phase gave way to what is known as the 'sub-Atlantic' phase, with blustery winds and cool, wet summers. As always with a major change of climate, an age died and another was born. The Bronze Age gave place to the Iron Age. Never again was Britain to enjoy so long a period of settled peace.

This, in fact, is one of the great turning points of history. The change in climate was pretty well world-wide. It set people moving everywhere, for those in the north who were dependent on dry warm summers for their harvest could no longer support life when the summers became cool and wet. Threatened with famine, they moved. The whole of Europe began to move. The peoples in the north and east pressed southwards and westwards in search of better lands, driving those already settled before them. And these movements continued long after the climate had settled down into its new rhythm. Western Europe was overrun again and again by wandering peoples from the north and east. Indeed, it may be said that the Roman Empire was born of the pressures brought about by this climatic change.

The first of these dispossessed peoples to reach Britain were men of the Bronze Age, the Urn folk, driven from their lands in north Germany by invaders from the north. They were followed by the Deverel-Rimbury people, who seem to have originated in central Europe. Both these peoples were akin to those already settled in Britain and, though there must have been some skirmishing, they seem to have been fairly easily absorbed. But soon, about 500 B.C., they were followed by people of very different temper: Celts, who had learned the use of iron and who came as conquerors. Wave after wave of iron-using Celts poured into Britain; some from across the North Sea, more from across the Channel. The Hallstatt Celts conquered southern England. It is to them we owe the great hill-forts that crown so many of our chalk hill-tops. They were thrown up, not as a protection against the resident Bronze Age population (which does not appear to have offered any really serious resistance), but as defences against the inroads of later invaders, the La Tène Celts. And they were highly successful. The La Tène Celts managed to settle in a few places along the south coast, but they were unable to make a conquest. Defeated here, they conquered and settled East Anglia, Lincolnshire and eastern Yorkshire, and from these bases they colonized a substantial part of Scotland. But the two most important Celtic invasions were those of the Belgae: the first about 75 B.C. and the second about 50 B.C. These two were invasions of Gauls, driven from France by the Romans. The first was followed by the two punitive expeditions of Julius Caesar: the second, when it had become established and powerful, by the Roman conquest under Claudius. We have arrived at historic times and at a water-shed in the story of British wild life.

What was Britain like during the sub-Atlantic phase, which began about 700 B.C. and lasted until about A.D. 200? A. G. Tansley, in his *The British Isles and their Vegetation*, describes it thus:

> While there was fairly extensive cultivation on the chalk and also on loam soils of the south-east, most of the English lowlands, for example the Weald and the Midland Plain, were covered with oak forest, mainly uninhabited and harbouring wolf, lynx, and bear, besides numerous deer. Oak forest also

> occupied the sides of the valleys in the hill and mountain regions of the west and north, giving place to pine and birch woods at higher levels and on the poorer and more sandy soils. In the extreme north of Scotland there was little or no oak even at low altitudes, though oak forest filled the bottoms and lined the lower slopes of the larger glens in the Central Highlands. In the limestone regions, for example on the Mountain limestone of the Northern and Southern Pennines, ashwood, the remains of which may still be seen in the Derbyshire dales, probably covered the sides of the valleys.... Throughout the lowlands water-logged areas were studded with many small meres surrounded by marsh or by fen, partly covered with willow and alderwood ... the most extensive (now drained and cultivated) occupying the sites of old estuaries from which the sea retreated and the salt was washed out by the rain. Towards the sea these estuarine fens pass into saltmarsh where the tides still have access. When the alkaline fen peat, fed by lime-containing ground water, has grown up above the winter water level, bog plants, largely bog-moss (*Sphagnum*), began to replace the fen vegetation and form bog – or mosspeat which receives the water directly from the abundant rainfall of a wet climate and is kept moist by the damp air ... such *raised bogs*, their surface slightly convex and above the level of the surrounding fen, remain in the central plain of Ireland, though in England most have been drained and destroyed. Bogs forming the same kind of peat occupy wide moorland areas not much above sea level in the wettest climates of all, i.e. in western Scotland and western Ireland, and these are called blanket bogs because they cover the country like a blanket. Blanket bogs also spread over high-lying plateaux where the drainage is poor and the rainfall heavy, as on Dartmoor, the Pennine plateaux, and in the Scottish Highlands. All these features of still existing fen and bog vegetation probably took shape at the beginning of the Sub-Atlantic period.

Such was the Britain into which the Iron Age Celts penetrated. They took over the lands, the open uplands and the cleared valley lands, of the Bronze Age peoples they subdued or drove out:

and, of course, they cleared more of the lowland loamy soils. But perhaps the most important effect of the first Celtic invasions was that they drove a substantial proportion of the Bronze Age peoples of southern Britain, East Anglia, Lincolnshire and Yorkshire westwards. These refugees naturally took with them what they could save of their herds and flocks, and so brought agriculture to western regions which before had not been considered worth the trouble of clearing. Many of the Bronze Age finds in western Scotland and north and central Wales, though from their nature dated prior to 500 B.C., must properly belong to a later date.

The Celts themselves were good farmers, good agriculturists and good pastoralists. The earliest Celtic invaders brought with them a light two-ox plough. Many of the hill-forts of southern Britain are situated in the midst of lands which still show traces of the small rectangular Celtic fields, which would have been cultivated with these light ploughs. (These, of course, were the lands which Bronze Age man had had to vacate when the climate became too dry to allow of their cultivation. The change of climate brought cereal growing back to the uplands.) The light two-ox plough could have been drawn by the comparatively small cattle then domesticated in Britain. The much heavier plough, with a broad-bladed share and a long, stout, concave coulter set so as to turn the furrow slice over, which was introduced by the Belgae, certainly could not. The Belgae must, therefore, have brought with them a larger, heavier breed of cattle. And it was the Belgae who introduced the long plough-strip. This is still to be seen in the form of 'lynchets' on chalk hillsides here and there in southern Britain. These lynchets must surely have been formed by ploughing parallel to the contours of the hill with the heavy plough and turning the spit downhill. Bronze Age man seems to have grown only flax and wheat of the type known as 'emmer'. The Belgae also grew barley, millet, oats and rye. Whether they had learned to alternate these crops, and to alternate them, or one or more of them, with fallow, is not known. But they grew them in sufficient quantities to establish an export trade with the Continent.

In addition to a larger and heavier breed of cattle, the Celts brought with them a larger, swifter, and distinctly more powerful type of horse. It was still pony-sized, but it was an animal more the size of the modern Highland pony (which is not to say that it looked like one: the modern Highland pony has had a considerable infusion of Arab blood) rather than that of the modern Exmoor pony: that is, a little over fourteen hands rather than a little under twelve hands. Though we have no direct evidence for this, it may be taken as virtually certain that they – not necessarily the Belgae, though they would certainly have brought their domestic animals with them – introduced an improved type of sheep and probably also an improved type of pig. The great increase in cereal growing would have been related directly to a great increase in the numbers of sheep and a more plentiful supply of dung. These flocks would need to be guarded and herded, and the Celts brought with them new types of dog.

Excavations at Maiden Castle have revealed the remains of a number of dogs – a few associated with human burials – and these have been dated to about 100–50 B.C. These remains, broadly speaking, fall into two types, both considerably larger than the Windmill Hill dog, but one very much bigger than the other. The smaller of the two, which is also much the more common, has been described by Professor Watson as a 'medium-sized, powerful, rather thick-set animal agreeing in its proportions and general character more closely with the chow than with any other breed with which I was able to compare it'. There would seem, therefore, to be little doubt that this smaller variety was a Spitz-type. And this, indeed, is very much what one would expect, for the Spitz family is one of the basic, one of the original, families of the domestic dog: one of the four families with counterparts in the wild pariah, from which, surely, the domestic dog is derived. This smaller variety would have been used for guarding, and also perhaps for herding, the flocks.

The larger type was, of course, the mastiff, the breed which so astonished the Romans when they first came to Britain. J. H. Walsh (*Stonehenge*) maintained that the mastiff is indigenous to Britain. It is not. The Romans were not (as Walsh thought) astonished at seeing a new breed of dog. They already knew the mastiff, under the name *molossus*, very well indeed – for they used it regularly in their gladiatorial shows – and they immediately recognized these British dogs for what they were, giving them the name *Pugnaces molossi.* What did astonish them was their size and ferocity. These British mastiffs were even larger and more powerful than the fierce fighting dogs used in the gladiatorial arenas, and forthwith some were sent back for the amaze and delight of the populace. Gratius, whose *Cynegetica* was written at the very beginning of the Christian era (probably before A.D. 8), recorded that the fierce molossian dogs of Epirus, which hitherto had been considered the greatest fighting dogs in the world, were speedily overpowered when pitted against the *pugnaces* from Britain. And from that time until the end of the Roman occupation a steady supply of *Pugnaces molossi* was sent from Britain to fight in amphitheatres throughout the Empire.

That British mastiffs were highly prized by the Romans is beyond question, but there is no truth in the story, which seems to have originated with Strutt and which is constantly being repeated even today, that the Roman Emperors maintained at Winchester an officer, with the title Procurator Cynegii, whose sole duty was to select and export fighting mastiffs. The legend would seem to be based upon a confusion between two words, *gynaecii* and *cynegii.* It is true, as Roman records prove, that there was stationed at Winchester an officer with the title Procurator Gynaecii in Britannis Ventensis. But this officer had nothing officially to do with dogs. He was the Administrator of the Imperial Weaving Works, a very important post. Had an officer been appointed to select and export mastiffs to fight in the gladiatorial arenas, his title would have been Procurator Pugnacium Molossorum.

We do not know when the mastiff first came to Britain, but we do know that it was not present during the Bronze Age proper. It has been suggested that it was first brought to Britain by the Phoenician merchant-venturers and exchanged, together with other commodities, for tin. It is an intriguing theory, for we know that the mastiff was very highly regarded in Assyria in the sixth century B.C. and it is not a far cry from Assyria to the Phoenician ports of Tyre and Sidon. Unfortunately, there is no firm evidence that the Phoenicians ever had direct contact with Britain. Modern archaeological opinion, indeed, is that they had not. But it is known that the Celtic tribes of central Europe used mastiffs in battle, so it would seem probable that it was the Hallstatt Celts who introduced the breed. It could not have been much later than that anyway – it could not, for example, have been the Belgae, for the Romans would then have known of the breed. And we know – it is the one certain historical fact – that when the Romans first invaded Britain they found here a breed of mastiff of a very superior sort new to them.

These great fighting dogs would, of course, have been used as household guards and, in time of need, also in the defence of the hill-forts. But their strength and ferocity would, surely, also have been used in hunting the bear. It is from this time that one may date the decline of the bear in Britain.

Though no identifiable remains have yet been found, there can be no doubt that the Celts also had other types of dogs. There is no reason to suppose that they had not a genuine herder – for the sheepdog family is one of the four original families of the domestic dog, and herders were in use long before this time in the Middle East – and it is certain that they had hounds. It is unlikely that they had scenting hounds, for these would have been unsuitable to their way of life, but it is beyond doubt that they had running dogs, hounds of the greyhound family, the fourth of the original families. Strabo (*c.* 64 B.C.–A.D. 19) tells us that one type of British hunting dog was particularly esteemed by the Romans on the Continent for its speed and strength. Sir Cyril Fox firmly believed that this was the breed now known as the Irish wolfhound, of which a fine Romano-British representation in bronze was found at Lydney in Gloucestershire. Indirect support for his belief is to be found in Julius Caesar's statement that the Britons did not eat the flesh of hares, despite the fact that the island abounded with them, adding that this abstinence sprang from a religious belief. It may be taken, therefore, that the hare was not hunted at all. But deer were hunted. The Celts had a great fondness for venison and wine. For the pursuit of deer they would have required a larger hunting dog than one capable of dealing with a hare: a larger dog than a greyhound. In any case the greyhound was already as well known to the Romans as it had been to the ancient Egyptians, and Strabo clearly implies that this was a type of hound new to the Romans. It may well, indeed, have been a prototype Irish wolfhound.

All this may appear to be somewhat irrelevant. It is not. What I have been outlining is the gradual creation of a rural civilization. The Celts brought the uplands and the lighter valley lands cleared by Bronze Age man, and the additional lands, all of them lightish soils, under relatively close cultivation. They built up great herds of cattle and great flocks of sheep, and they must have improved the standard of both, and no doubt also of their pigs, greatly, for they were certainly not ignorant of the principles of selective breeding and the benefits derived from them. The proof of this lies in the mastiff and the swift and powerful hound (whatever it may have been). These breeds were exported, which means that the qualities for which they were famous must have been maintained and that, in turn, means that breeding must have been rigorously controlled by men who knew well what they were about. If they could do that with their dogs, there is no reason to suppose that they did not also do it with their cattle and their sheep and their pigs.

All this must, of course, have led to a marked increase in the numbers of, and probably also in the variety of, seed-eating and insect-eating birds with a natural addiction to open country – the increase in the numbers of domestic animals would, inevitably, have led to a great increase in the numbers of insects associated with them and with their droppings – and consequently to an increase in the numbers of their natural predators. Obviously, too, these farmers would have taken such steps as lay within their power to protect their herds and flocks, and to reduce so far as possible the numbers of the really dangerous predators. And with their new types of dogs these would have been pretty effective measures. Life must have become, at least so far as the cleared country was concerned, much more difficult and hazardous for the bear, the lynx, and the wolf. But there was no question of extinction for any of these species. There was always the deep oak forest, the vast areas of heavily wooded waterlogged land, to serve as sanctuary for the harassed; to provide a safe refuge for breeding purposes. The Celts did not attempt the clearance of these areas: probably rarely penetrated them and then only in pursuit of some wounded beast. Nor did the Romans attempt their clearance.

Now, there must have been some reason for this. So far as Bronze Age man is concerned, the reason is evident. Quite apart from the fact that there was no pressure on living space – that there was more than sufficient light land available to meet the needs of the population – he simply had not the implements necessary to bring heavy wet land under cultivation. But the men of the late Iron Age, the Belgae, did have the necessary implements: and so did the Romans. Their heavy plough was the same as that used by the Saxons, who did make small clearances and establish small settlements in these wet forest lands. Why, then, did not the Belgae or the Romans?

For the Belgae there was still no real pressure on living space. They had no need to do so in order to support their increased population. Nevertheless, the presence of great tracts of dense forest lying between one upland region and another must have been most inconvenient. The Belgae had the ability to clear at least portions of these forests and to establish small settlements, and this would have been of great help to communications between one upland region and another. They did not do so. Why? The obvious reason would seem to be that the dense forests were the natural home of bear, lynx, and wolf and that, protected by thick cover, these beasts were the more easily able to attack man and his womenfolk and his farm animals. Man, as hunter, would venture into the wet forest lands, but settlement was too hazardous to be worth attempting. At first sight this reason would seem to be even more cogent for the Romans, for the Roman civilization was essentially an urban civilization. They were primarily townsfolk and might, therefore, be expected to be the more scared of wild beasts. But man has never been deterred by such hazards as these. There must have been more to it than that.

It has been suggested that the real reason was disease: disease not only of man, but also of his domestic animals. In his *The Personality of Britain*, Sir Cyril Fox put forward the theory that early man 'learnt by experience that if his sheep strayed for any length of time into such lands, they sickened and died of liver-rot, a disease known today to be most prevalent in low wet grounds. He himself, doubtless, under similar conditions, suffered from ague (malaria, due to the *Anopheles* mosquito, an inhabitant of stagnant pools) and from rheumatism, a disease of wet lands.'

No one would deny that liver-rot, which is caused by the liver-fluke (a flatworm), is most prevalent in low wet grounds and no one would deny for a moment that it can cause most serious losses; not only in sheep, but also, though more rarely, in cattle. Indeed, until quite recently the annual loss from liver-rot, in Britain alone, was estimated at around one million animals. It is, so far as sheep are concerned, a most intriguing theory. But it is, I think, one that must be dismissed. Sheep are not forest animals. Before they could graze the wet lowlands the forest, or at least parts of the forest, would have to be cleared. And this, we know, was not done. It is, I think, stretching a very long bow to maintain that early man would relate the death of a stray sheep to a short sojourn in the wet forest lands and that he would, because of this, leave the wet forest lands alone. Quite apart from that, few sheep straying into the dense wet forests of that time would have had much chance of getting out again alive. No, I think that we must dismiss that theory.

But malaria is quite another matter. Today we regard malaria as primarily a tropical or sub-tropical disease. It is not generally realized that it was prevalent, particularly in the fen districts of East Anglia and Sedgemoor, until the middle of the nineteenth century. How much more so must it have been in the early years of our history, before extensive drainage was carried out! Here, surely, there can be little doubt. Man, entering the forest, making a small clearing by some pool with a view to settlement, would get the ague and maybe he, or his wife or one of his children would die. The direct relationship between cause and effect would be all too evident in due course; would be quite sufficient to cause the wet forest lands to be given as wide a berth as possible.

Given a wide berth they were. Consider the history of the Weald. This huge area, about one hundred and twenty miles in length and some forty miles wide, was the 'Sylva Anderida' of the Romans, the 'Andredsweald' of the Saxons. (Weald and wold, by the way, are same word, coming from the Welsh *coed*, meaning 'a wood'.) The Romans drove a few roads, remarkably few, but notably Stane Street, through Sylva Anderida. These roads served commercial rather than military purposes. They seem primarily to have been built to link iron-working sites, of which there are about ten (in the whole area of the Weald) which can be dated with certainty as Roman. Except for these iron-working sites, and they were probably worked only for a few weeks in summer, there is no evidence of Roman settlement in the Weald proper. It remained, throughout the Roman occupation, a land apart. And with the collapse of Roman rule it reverted to the damp isolation that was its in prehistoric days.

At the end of the Dark Ages, some two hundred years after the withdrawal of the Roman legions, the Saxon kingdoms began to emerge into the light of documented history. And some time after this – no one knows when; but it may well have been a century or two – they began to make small enclaves in the Weald: but very few, as is clearly indicated by place-names. The names that matter are 'ing', 'den', 'fold', and 'field'. A 'den' was a woodland pasture. You will find a number of place-names ending in 'den' in the Kentish part of the Weald. 'Ing' refers to the place where the followers or relatives of a named man lived: for example, Steyning= the place of Stane's people, Worthing=the place of Wurdd's people. If you plot the 'ing' names on a map, you will find that they all tend to fall on the borders of the Weald rather than within the Weald proper: not one occurs on the true Wealden clay. The word 'fold' means an enclosure in woodland. The 'fold' was the first small clearing man made in the forest, his first attempt at settlement; perhaps by charcoal-burners. There are a number of 'fold' endings in the western Weald – for example, Chiddingfold (the 'fold' of Cidda's people) – and some of these are on the Wealden clay. More enlightening still is the termination 'field'. Nowadays, when we speak of a field, we mean an enclosure, a fenced-off open space. But this is a new development: the modern meaning came with the Enclosure Acts. Until about two hundred years ago the word had exactly the opposite meaning. Then it meant open, unfenced land free from trees. The field was the open land at the edge of the woodland. And the first clearances made by man would, of course, have been along the edges of the woodland. Thus, by plotting the situations of place-names terminating in 'field', one can trace the boundaries, and so the extent, of ancient forests. Consider, for example, the great number of place-names with this termination in the neighbourhood of Windsor Great Park; Winkfield, Warfield, Binfield, Arborfield, Swallowfield, Heckfield. Indeed, without any considerable gap, one can trace these 'field' names through Sherfield-on-Loddon to Englefield and on to Bradfield, high on the Berkshire Downs. It is instructive to trace this line out on the map. Here was once, bounded to the north by the Thames, a great forest. Similarly, you will find 'field' names scattered around the edge of the Weald, particularly along the southern edge, and it is instructive to note how close many of them are to places with the 'ing' termination. Evidently these 'fields' were the first attempt of the 'ing' peoples to win ground from the forest.

What all this amounts to is this: that the Saxons did make clearances along the edges of the forest (but not very many of them: the Weald was a tougher nut to crack than the great forest around Windsor), that they did make inroads, but not many, into the forest. They did not embark upon any systematic clearing of the forest: did not attempt any serious settlement, any large-scale cultivation. Indeed, it was not until late Elizabethan times, not until the end of the sixteenth century or the beginning of the seventeenth, that, following centuries of piecemeal clearing for timber or for charcoal to smelt the local iron ore, there was any serious attempt at cultivation on any scale. Even today the Weald is one of the most heavily wooded regions in Britain. Even today, standing on one of the higher hills on its borders, you have the illusion of looking out over a vast, an endless, forest.

Bearing in mind its defiance of man over the centuries, the Weald must surely have been one of the last strongholds in southern England, perhaps the last, of the wolf and the wild cat.

Chapter 4

ROMAN AND SAXON

THE ROMAN conquest of Britain is commonly believed to have been carried out for strategic purposes. Personally, I doubt if this was the sole reason. It would surely have been easier and considerably cheaper, both in terms of money and man-power, to form a stronger defensive frontier along the Channel coast. I believe that the conquest was undertaken for economic as well as strategic reasons. When the Romans occupied a land they expected to recover their costs and to reap rich rewards: they expected the occupied territory to pay rich dividends by way of raw materials and taxation. From this point of view Britain was an ideal country for conquest. It provided a new and expanding market for Roman trade: Roman goods and Roman money poured in. It was, moreover, a land rich in minerals, the most valued prize of conquest.

All minerals belonged to the State, and the Romans set about mining them straight away. The lead mines of Mendip were in production by A.D. 49, a mere six years after the initial invasion; and lead mines began to be worked in Derbyshire, Shropshire, North Wales and Yorkshire as soon as these regions were brought under firm control. Silver was extracted from this lead by special smelting and this accounts for the rapid development of the British lead mines: silver formed the chief Roman coinage. Tin was mined in Cornwall, copper in Shropshire and North Wales, iron in Northamptonshire, Lincolnshire and Yorkshire, the Forest of Dean and, as we have seen, in the Weald. Gold was mined at Dolau Cothi in Carmarthenshire and there was also some opencast coal mining, though there is no evidence that coal was exported. Salt mines were also worked and there was a considerable trade in jet and in pearls. Apart from all this, there was a large export trade in furs, in wool, and in corn. Britain was indeed an island worth conquering, paying rich dividends on the capital (no small sum) invested.

All this did not happen at once, of course. The native tribes, a pretty war-like lot, had first to be subjected. This was done by the establishment of garrisons at strategic points and by the building of a network of roads, which enabled troops to be moved quickly from one point to another should need arise. Once the country was under firm control these roads became a major commercial asset. Market centres grew up along them from which goods could be transported quickly and safely to the ports. These market centres were sited at recognized tribal centres and were at first no more than small groups of simple timber huts. But they were soon to develop into towns laid out on the approved Roman principle. The Romans were really only at home, only at ease, in a town. The Roman civilization, as I have said, was essentially an urban civilization.

Except for the building of roads, except for the draining of some of the marshes, the Romans left little mark on the countryside as such. Where land was already in cultivation by the native population, though there were probably some changes of tenure here and there, there does not seem to have been any real change in the system of cultivation. The farmers on the chalk uplands of the south of England, the farmers of the late Bronze Age and of the early Iron Age, the Hallstatt Celts and so on, had made small rectangular fields and this system was not altered under the Romans. Moreover, the people who tilled these small fields continued to live in small beehive thatched huts of the prehistoric pattern. The 'Roman villa' came much later; and was born in the town.

For the towns were used, not simply as garrison towns for the legions, not simply to house Romans, but to romanize the native population. The tribal nobility and gentry were encouraged to adopt Roman ways and to give their sons a Roman education. It is significant that in every Roman town in Britain it was the amenities – sewerage, public baths, a theatre, a town hall, comfortable private houses – that were built first. Fortification – the outer walls and the four gateways – came later. In the words of Dr Richmond:

'the political and cultural education of the native aristocracy was the first and fundamental condition of civilizing the province as a whole'. The 'Roman villa' was born of this deliberate policy. Members of the native aristocracy, wealthier members of the gentry, or their sons, having absorbed sufficient Roman culture, would return to their farmlands and set about building themselves the status symbol, a house in the Roman manner, built to a rectangular plan and with partitioned rooms. At first these houses were built of wood, but as wealth increased, as inevitably it did with increased trade and an increased population, more and more were built of stone and many had baths and central heating. Few of these Roman villas were, in fact, owned or occupied by Romans. They were the homes of romanized Britons. The Romans, for the most part, stuck to the towns.

But it is important to realize that the Romans did not move their troops around. There was no 'tour of duty' as has always been the custom in the British Army. The Roman soldier did not serve for a year or two in Britain, then have a spell of home duty, then serve for a year or two in Spain or Germany or north Africa or wherever, and so on. He stayed where he was posted. The legions and the cavalry remained in their stations for generations together, sometimes for centuries. And when, at the end of his service, the legionary retired, he did not, as a general rule (only the highest ranks would have been able to do so), return to his homeland. He remained in Britain, which had become his homeland: where, indeed, he, and for that matter his father and his grandfather, may well have been born. Four towns – Colchester, Gloucester, Lincoln and York – were actually built for the main purpose of housing army veterans, though they were also, of course, great administrative centres. Each of these towns had lands attached to it which were divided among the veterans and cultivated by them. The practice of long, virtually permanent service abroad meant, of course, constant intermarriage at all levels of society with the native population and this, in turn, meant the steady romanization culturally of the whole settled part of the country.

Moreover, the steady increase in population – it has been estimated that the total population of Britain at the time of the

Roman conquest was less than half a million, but that at the end of the occupation it was in excess of a million and a quarter – meant that more and more land was brought under cultivation, more and more cleared for pasture. Some of the villas were the homes of great landowners, farming estates of a thousand acres and more. On these large farms, and on others of more moderate size, the heavy coulter plough would have been used and the fields would have been correspondingly large. But along with the wealthy villas and the great estates were many small farms, many small-holdings, on which the more primitive plough was used. As time passed and more and more legionaries married into the native population, more and more of these small family farms sprang up away from the already settled areas. In the hill lands of the west and north particularly, marginal land was gradually cleared to form grazing for sheep and cattle. Though, now, we have not the slightest idea what the effect of all this on wild life was, there can be no doubt that it must have been considerable.

The Roman roads, too, undoubtedly had a great effect. These Roman roads were 'highways' in the strict sense of the word: that is, they were built high above the level of the surrounding countryside. J. D. Margary, in his *Roman Roads in Britain*, says that the minor roads were raised only one to two feet above the normal ground level on either side, but that the major roads were frequently four to five feet high. For some roads, at any rate, this would appear to be an under-estimate. Writing half a century earlier than Margary, Codrington, in his *The Roman Roads of Britain*, describes the road from Old Sarum to Bradbury as being raised six to seven feet. Daniel Defoe, in his *A Tour Thro' the whole island of Great Britain*, which was written during the first quarter of the eighteenth century, describes the Fosse Way as being raised seven to eight feet above the level of the adjoining countryside. It is certainly nothing like that height today, but that is not to say that it was not when Defoe rode along it. Defoe was a singularly careful and accurate observer. However, the actual height at the time of building does not matter: it would have varied from road to road and at different points on a road anyway. The point is that these great highways, some of them twenty-four feet wide, were built along the tops of embankments, some of them eight feet high, and perhaps more, in places. Sir Edward Salisbury, in his *Weeds and Aliens*, has pointed out that the sloping sides of these embankments would have provided a countrywide extent of open ground, probably exceeding six thousand acres in extent, eminently suitable for colonization by weeds and aliens. They would have done something else, too, We all know how rich in wild life, particularly in plant and insect life, are the embankments and cuttings of our railway system. Indeed, I have in recent years come to regard south-facing railway embankments as the last remaining refuge for many of our butterflies. The legions, tramping these highways in summer, must have marched above a wealth of colour, plant and butterfly, and to the music of many grasshoppers and crickets.

The Romans – and, of course, peoples before them; from Neolithic times onwards – must unwittingly have introduced many weeds. It is virtually impossible, if not quite impossible, for anyone to move over land without picking up and carrying seeds with him. Many of our common weeds, it is generally agreed, are of southern European origin. They were probably brought in on the clothes or sandals of colonists. The Romans are widely credited with the introduction in this way of, among others, pimpernel and poppy, the two sowthistles, fool's parsley and the small stinging nettle. But all of these may well have been brought in by the Celts. More certainly of Roman introduction is darnel, for Virgil (70–19 B.C.) refers to it in *The Georgics* as a pestilential weed in southern Italy in his day.

But the Romans also made deliberate introductions of plants. To them we owe the presence of ground elder: now regarded as a most pernicious weed, but in Roman times (and for centuries later) highly prized as a medicinal herb. To the Romans, too, we owe hemlock, another plant which they used medicinally, and also henbane, which was greatly valued for its narcotic properties.

The Romans may also, I think, be credited with the introduction of two birds: the domestic chicken and the pheasant. There is, it must be admitted, some doubt about both. Chicken bones are said to have been recovered from Bronze Age sites which, if they are indeed the bones of chicken, would give the domestic fowl a very long British history. There is no doubt about the dating of the sites: they are unquestionably Bronze Age. But there is a very considerable doubt about the identity of the bones and no certainty whatsoever that they are of the same age as the deposits in which they were found. And, with the possible exception of those found at the Glastonbury Lake Village, all chicken bones of supposedly Iron Age date must be viewed with the gravest suspicion. The Glastonbury Lake Village belongs to the very end of the Iron Age. It was probably established not earlier than 50 B.C.: that is, after the punitive expeditions of Julius Caesar and at a time when there was already well-established and frequent contact, commercial and otherwise, between Britain and Rome or, at any rate, Roman Gaul. It is certainly possible that the domestic fowl was introduced by the Glastonbury people before the Roman conquest. But it must be remembered that the Glastonbury Lake Village was occupied until well after the Roman invasion; indeed, until after Vespasian had pacified the west. No one can, with certainty, date the chicken bones found at the Glastonbury Lake Village as pre-Roman.

The pheasant also poses some problems. Pheasant bones are said to have been found in a Neolithic midden at Ardrossan on the coast of Ayrshire. There is absolutely no evidence that Neolithic man succeeded in domesticating the pheasant: so, if these bones were correctly identified, it must mean that the pheasant was present as a wild bird in Scotland a matter of four thousand years or so ago. The likelihood of that being so is so improbable that I find myself wholly unable to accept the identification of these bones as correct. I also find myself unable to accept the identification of bones found at a late Bronze Age site at Barton Mere in Suffolk as being those of the pheasant. Again, there is no evidence whatsoever that the men of the late Bronze Age (say, 1000–750 B.C.) domesticated the pheasant. If the pheasant was present as a wild bird in Suffolk at that time, then one would expect it to have been present elsewhere also. There is no evidence that it was: and no pheasant bones, so far as I am aware, have been recovered from Iron Age sites. This, too, must surely be a mistaken identification. Pheasant bones have also been recovered from Roman sites at Silchester and Corbridge. James Fisher – and it would be a foolish man who would argue with so great an authority – has dismissed the Silchester bones as being wrongly identified. But he accepts the Corbridge identification as valid. So here we have the first certain evidence of the presence of the pheasant in Britain. The Romans were, we know, enthusiastic keepers of ornamental pheasants and I think it may be taken as beyond question that it was they who first introduced the bird to Britain. But not as wild birds: their pheasants were domesticated and were kept in pens. There is no evidence that they ever escaped to establish themselves in the wild: no evidence, indeed, that they survived, as domestic birds, beyond the end of the Roman occupation.

Some authorities believe that the hare referred to by Julius Caesar was the blue or mountain hare (*Lepus timidus*) and that the brown hare, which is the common hare of Britain, was introduced at some time during the Roman occupation. I do not know upon what evidence this belief is based. I like to think that our brown hare is indigenous.

But the fallow deer certainly is not indigenous. Its introduction has been variously attributed to the Phoenicians, to the Celts, and to the Romans. The Phoenicians can be eliminated straight away. It is, surely, ridiculous to imagine that thrifty merchant-venturers, making the long voyage from the Mediterranean, would have taken up valuable deck-space with a cargo of deer (and the food and water necessary to keep them alive) on the off-chance of selling it to the people of an island already well stocked with deer. Much the same argument applies to the Celts. The only Celts who could have done so were the Belgae. But the Belgae came as conquerors. It is true that they did establish a kingdom in south-eastern England and that they maintained close contact with their fellows across the Channel. But from the time of the first Belgic invasion in 75 B.C. to the second under Commius in 50 B.C. there was almost continual tribal warfare. And this continued, small tribal wars developing into major tribal wars – the kingdom of Cunobelin was founded by the subjugation of neighbouring tribes and maintained by force of arms – until the Roman invasion of A.D. 43. An animal such as the fallow deer could have been introduced only in a time of settled peace. So we are left with the Romans. Now, there is no direct evidence that the fallow deer was introduced by the Romans. But we know from the evidence of Domesday Book that it was then well established and widespread. It could not have been brought in during the Dark Ages following the Roman withdrawal and the Saxons and the Danes were far too busy – and as peoples were not so minded – to bother about the introduction of another species of deer. It must have been the Romans and the introduction must have occurred at a time when southern England was completely pacified and prosperous. It is easy to imagine the owner of a luxurious villa, the owner of a great estate, wishing to see white-spotted deer at liberty on his property. Throughout our history wealthy landowners have always had such fancies. In general, our fauna is the richer for them.

The aurochs, though it survived in parts of mainland Europe until the late fifteenth or early sixteenth century, was extinct in southern England before Roman times. It is impossible to believe that Julius Caesar, who noted the presence of so small an animal as the hare, could have remained ignorant of the presence of so large a beast as the aurochs had it existed. But the brown bear was still present in some numbers, for the poet Martial (A.D. 40–104) tells us that bears were exported from Britain to take part in the gladiatorial shows, where they were held 'in great admiration'. There was, moreover, in Roman times an export trade in British bear skins. The bear must, therefore, have been fairly common, but we have no idea of its distribution. It is not impossible that it was still present in the Weald during the early years of the Roman occupation.

One further aspect of the Roman occupation, the drainage of marshes, must be considered here. The Romans were well aware of the value of marshland, of the fertility of the soil for the growing of crops, and had drained considerable areas in Italy and elsewhere in southern Europe. They were not the sort of people to neglect such land in Britain. They carried out extensive drainage of the Fens by a system of wide canals, the most notable being the Cambridgeshire and Lincolnshire Car Dykes, the latter over seventy miles long. The land so drained was brought into cultivation. Presumably, some drainage work was also carried out elsewhere. One cannot, for example, imagine the Romans overlooking the worth of such rich soil as that of the Vale of Avalon; and there is some evidence that they built sea-walls as a first step to the deliberate exploitation of this area. This interference with the natural conditions of flood and swamp may well have been the reason for the abandonment by the Celts of the Glastonbury and Meare lake villages. Such drainage work must, inevitably, have had a considerable effect on the marsh-dwelling, water-loving birds; increasing the numbers of some species, decreasing the numbers of others. With one exception, we have no knowledge how individual species were affected. The one exception is the grey or Dalmatian pelican. This bird was resident and breeding in the Vale of Avalon, and in some numbers, around the time of the Roman invasion and almost certainly it was also present in the great fenland around the Wash at that time, for fossils have been found in Cambridgeshire and in Norfolk. There seems little reason to doubt that the extinction of the Dalmatian pelican as a British species was due, indirectly, to Roman engineers.

A word must also be said about dogs, since the use of hunting-dogs has a considerable effect (and in more ways than the obvious one of killing) on wild life, and it is evident that the Romano-British country gentleman was partial to the chase. In addition to the mastiff and the wolfhound, we know from Oppian's *Cynegetica*, the four volumes of which appeared between A.D. 211 and 217, that a greyhound – strangely enough, described as having blue eyes – from northern Britain was highly prized for its speed and courage, and that there were at that time at least two sorts of scenting hounds in use in Britain. His description of the first of these resembles the description given by Xenophon of the hounds he used in Greece in the third century B.C. so closely that there can be little doubt that it was introduced by the Romans. But his description of the other—

> By painted Britons brave in war they're bred,
> Are Beagles called, and to the chase are led;
> Their bodies small and of so mean a shape,
> You'ld think them curs that under tables creep—

suggests that it was present in Britain before Julius Caesar's first excursion and may, therefore, be indigenous. Scholars may question the rendering of Oppian's *Agassoei* as 'Beagles', but the subsequent lines of the poem leave no doubt that he was alluding to diminutive hounds. However, we need not rest on Oppian. The word 'beagle' comes from the Celtic *beag* (the modern Welsh *bach*), meaning 'small'. It is the 'brach' of medieval writers and of Shakespeare also. Moreover, we know from Canute's forest laws (which, incidentally, were much more savage than the famous forest laws of William the Conqueror) that beagles were in common use in Saxon times, for 'small hounds' were specifically excluded from mutilation. The presumption, of course, is that then as now these small hounds were used to hunt the hare. I think it more likely that they were used to drive foxes into nets: a common practice until the birth of foxhunting proper.

One more interesting sidelight on Romano-British life of the time is to be gained from Oppian's work. The aristocracy had not only their great estates and their luxurious villas with their mosaic pavements, their baths and their central heating: they also had their lap-dogs. From the description given by Oppian we get the picture of a tiny 'spaniel': a fluffy-haired creature, perhaps about the size of the modern King Charles spaniel. A people so civilized that they were prepared to give house-room to and lavish care upon so useless an animal would be just the sort of people to introduce a white-spotted deer as the ultimate status symbol.

With the withdrawal of the Roman legions in 410 Britain fell into decay. Actually the process of decay had started much earlier. Few of the great villas, which in many areas had formed the basis of Romano-British rural economy, survived the great barbarian raids of 368 and subsequent raids wrought further destruction to that economy. But the withdrawal of the legions meant the complete breakdown of central administration. The roads were no longer kept in repair and long stretches of them soon became unusable, canal banks were no longer maintained and large areas of drained land reverted to swamp, and many of the towns – for example, Silchester, Caister, Verulam – were deserted. The Saxon invaders found a countryside that could not, in essentials, have differed radically from that found by the Roman invaders four centuries earlier: a land with great stretches of continuous woodland and large areas of swamp, in which, with the exception of the open chalk and limestone uplands and some of their valleys, there were but few, relatively small islands of cultivation. We know virtually nothing about British history during the Dark Ages and nothing at all about the wild life of the time. But it is a safe bet that, during the four centuries immediately following the Roman withdrawal, predators such as the wolf greatly increased in numbers and in range, and that others, such as the bear and the lynx, probably succeeded in re-establishing themselves in areas from which they had been driven during the Roman occupation.

Our knowledge of wild life during Saxon times is really no greater. But something can be deduced from our knowledge of the Saxons themselves. They were, when they arrived in Britain and for a long time afterwards, a people wholly unaccustomed to town life. Deserted towns, such as Silchester and Verulam, were left deserted. And even in places, such as York and London, where almost certainly there had been continuous occupation since Roman times, the Saxon settlements grew up beside, and not in, the Romano-British town. One reason at least for this was that the Saxons did not know how to build in stone and had no idea how to repair derelict stone buildings – after their conversion to Christianity, the early church builders had to import masons from abroad – and preferred to live in timber or lath and plaster buildings. Saxon town life, as opposed to Saxon settlement around the fringes of an existing Romano-British town, really did not commence until about the ninth century and was then virtually confined to trading settlements on harbours and navigable rivers, except for the 'burghal system' (fortified centres built for the protection of a tract of country against Viking raids and which were to be the boroughs of a later age) inaugurated by Alfred the Great. There would seem to have been some difficulty in persuading a rural-minded people to settle in these boroughs, for each borough was given arable fields to be shared among the inhabitants and common pastures for the grazing of their sheep and cattle, and most of the houses seem also to have possessed vegetable gardens. As the sheep and cattle often grazed within the walls, the average Anglo-Saxon town must have had a distinctly rural aspect and flavour.

Basically, the Saxon economy was an agricultural economy. And they were good farmers. Wheat and barley were the principal crops, the latter being used not only for bread but also to provide malt for brewing – the Saxons were great drinkers and great keepers of bees, for honey was not only their sole means of sweetening, but also the chief ingredient of mead, one of their favourite drinks – but rye, beans, peas and flax were also grown. All the available arable was not cultivated in any year: part, a half or a third, was left fallow, and the part under cultivation was further divided among winter-sown and spring-sown crops. To quote C. S. Orwin: 'autumn sowing on the fallow was followed by winter cultivation of the stubble and spring sowing, after which fallow cultivation and hay-making occupied the farmer until the corn harvest finished the farming year'. This was not a system that provided an easy livelihood. It required very hard work all the year round to produce even a moderate living. A bad year meant scanty fare: a succession of bad years, famine. How difficult life was during this period may be deduced from the fact that the estimated population of one and a quarter million at the end of the Roman occupation had, on the evidence of Domesday Book, increased not at all six hundred years later. It was this – the ever-present possibility of a bad year, not pressure on the available living-space – that prompted the Saxons to win more and more land from the woodland and the waste. Land so won had to be protected: the crops from the ravages of deer, the livestock from attack by predators. It was the incessant war waged upon deer by farmer and peasant, not merely for the protection of crops but also because in winter deer were virtually the only source of meat, that gave rise to the harsh forest laws of Canute.

As to predators: wolves at this time must have been so numerous as to be a menace. Hywel Dda decreed that 'a wolf and a fox and various others which do nothing save mischief and on which no legal worth is fixed, it is free to all to slay them'. The skin of a wolf was valued at eightpence: a considerable sum in terms of today's values. William of Malmesbury, Caradog of Llancarfan, and, much later, Holinshed all state that the Saxon King Edgar in 962 imposed a tribute of three hundred wolf skins yearly on the Welsh. Caradog of Llancarfan was one of the company of men of letters (Geoffrey of Monmouth was another) who gathered around Robert, Earl of Gloucester, the illegitimate son of Henry I. The date of his birth is unknown, but it is thought that he died in 1147. He was, therefore, writing long after the events he describes. He says that King Edgar

> went into Gwynedd, and causing Iago, son of Idwald, to come to him, he imposed upon him a tribute of three hundred wolves yearly, in place of that due to him by the old law, with liberty to kill them in any place they pleased throughout the whole island of Britain; and so peace was established in Gwynedd. And that tribute was paid to Gwynedd for more than five and forty years, until not a wolf could be found in the whole island of Britain. After that the King of the Saxons commuted the tribute for gold, silver and cattle, as formerly.

All the chroniclers, presumably deriving their information from Caradog, agree upon the amount of the tribute, namely three hundred wolves a year, but there is a wide discrepancy as to the length of time during which the tribute was paid. Some say four years; some only two. Caradog's figure of forty-five years is obviously inaccurate. King Edgar died in 975. Forty-five years would have carried the tribute through the reign of Edward the Younger and through most of the reign of Ethelred the Unready. Edward the Younger was assassinated – proof enough of a disturbed kingdom – and Ethelred was not the sort of monarch to whom the unruly Welsh would have been likely to pay attention, let alone pay tribute. Edgar, it is true, was a strong king; but there is absolutely no shred of evidence that he was able to impose tribute upon the princes of any part of Wales and certainly not upon the prince of so distant and mountainous a part as Gwynedd. What would seem to have happened is that he hired a professional wolf-hunter from the Prince of Gwynedd and that this man was contracted to kill a minimum of three hundred wolves a year. It seems likely that, once hired, this man would have remained in the king's service for the remaining thirteen years or so of his reign. That would give a total of some 3,900 wolves destroyed under contract. All the chroniclers are agreed that when the tribute ceased 'not a wolf could be found in the whole island of

Britain'. This we know to be wildly inaccurate. But, in a fairly restricted area, a toll of some 4,000 wolves would have made a noticeable difference. Indeed, it does not appear improbable that wolves were exterminated in some part of southern England – say, in the vicinity of Winchester – at that time, for Iago, a professional at his job, permitted to kill where he pleased, would have been likely to conduct his activities where they would attract the most attention and appreciation.

The brown bear, plentiful in Roman times, had certainly disappeared from the south of England by Saxon times. No one knows exactly when the bear became extinct in Britain. A description of bear hunting is given in *Y Naw Helwriaeth* ('The Nine Huntings'). This is generally accepted by scholars, on internal evidence, as a very ancient Welsh text, but the actual date is not known. The earliest known MS., which is in the National Library of Wales, dates only from about 1560. The text may be accepted as evidence for the existence of the bear in Wales at some time, but no more than that. It is significant, however, that the bear is not mentioned in the laws of the Welsh king, Hywel Dda. So large an animal, had it existed, could not possibly have been overlooked by so careful and thorough a man. It may be taken, therefore, that it was already at that time (948) extinct in Wales. But the bear is represented on tenth-century 'hog-back' stones in Durham and Yorkshire, which suggests that it was then present in the north of England. It is generally thought to have become extinct before the end of the tenth century in the north of England, but it may well have survived into the eleventh century in parts of Scotland.

Again, no one knows the date of the extinction of the lynx in Britain. Sir Cyril Fox, a cautious man, considered it possible that it may have survived into early historic times. Personally, I should regard this as certain. Indeed, I should have thought a considerably later date probable, if for no other reason than that it did not become extinct in France until well into the nineteenth century. Bearing in mind its size and the nature of its habitat, I can see no reason why the lynx should not have survived at least as long as the bear.

The aurochs we know, on such scant evidence as we have, to have been extinct in southern England before the arrival of Julius

Caesar. And we have no information as to its presence anywhere in England in Roman times. It seems to have become extinct in northern Scotland about the middle of the ninth century, but there is a possibility that it may have survived in the wilder parts of Wales until the middle of the tenth century. There is no direct evidence to support this assumption, but it may be significant that there are a number of references to *corn bual* (buffalo horn) in the laws of Hywel Dda, and that in these laws *bual* is never used to denote domestic cattle. It is evident that wild cattle of a type different from the ordinary domestic cattle were then present in Wales. I am well aware, of course, that this is not evidence for the presence of the true aurochs. But it is not beyond the bounds of possibility that true aurochs blood ran in the veins of these wild cattle. It is difficult to account for them otherwise. We shall come across these mysterious animals again.

Of the rest of the wild life of Britain in Saxon times we really know very little; and that chiefly concerns birds. The corpus of Anglo-Saxon literature surviving is very small. The authors were scholars, not naturalists. They were interested in religion, in history, in the things of their mind: their references to birds are for the most part casual and passing. This is not to say that these men did not know their birds – it is evident that most of them did – but that they were not concerned to list all they saw. So it is not, perhaps, surprising to find that, at the end of the Saxon period, we have only seventy-five species recorded in the literature. That figure is not, of course, to be taken seriously. Most of our present common birds were probably in Britain then – the robin, the chaffinch, the goldfinch, the hedge-sparrow, the song thrush, the starling, the crow and others are specifically mentioned, for example – but their distribution and numbers would have been very different. And probably the summer migrants were not so very different, if different at all: the nightingale, for example, is mentioned before the end of the seventh century. But there would then also have been species common in certain types of habitat, which are now scarce and local in those habitats: for example, the capercaillie (which actually became extinct as a British bird about the middle of the eighteenth century, but which was reintroduced in 1837) and the crested tit in the pine forests, and the black grouse in the great birchwoods, of the north. And there would then have been nesting in Britain some species which now come to us only as winter visitors. I think that there can be little doubt that the brambling, the fieldfare and the redwing were once regular nesters in northern Britain and that they ceased to nest only with the disappearance of the great birchwoods; which means that they would have been resident British birds in Saxon times and, indeed, until a good deal later. The great auk, now wholly extinct, must then have been common on the cliffs of all our northern coasts, even though it is not recorded in Anglo-Saxon literature. But the Saxons do not seem to have been very interested in sea birds. There is no record in the surviving literature of guillemot or razorbill or puffin (birds which must have been as common on cliffs then as they are now; and which, perhaps, are ignored for that very reason) and only two gulls are mentioned: the kittiwake (recognized because of its distinctive cry) and one other, which was probably the herring gull. But both the mute swan and the whooper swan are mentioned and the crane is distinguished from the heron.

If the Saxons do not seem to have been interested in sea birds, they were certainly most interested in the birds of prey. The white-tailed eagle (the sea eagle or erne) and the golden eagle, the asprey and the peregrine, the kite and the buzzard, the goshawk and the sparrow-hawk, the kestrel and the tawny owl are all mentioned. And there are some real surprises: the gyr-falcon and the eagle owl, for example. The gyr-falcon would have been the bird we know today as the Greenland falcon or as the Iceland falcon. These are races rather than genuine sub-species of the typical Norwegian gyr-falcon. Both races nest in Iceland and the Iceland falcon also in the Faroes. Both occur in Britain as vagrants in winter: the Greenland falcon annually in very small numbers (perhaps two or three in a normal winter), the Iceland falcon much more rarely. The eagle owl, by far the largest European owl, nests in Norway and also in Spain. It is a very rare visitor to Britain in winter. Very rare indeed: I am pretty sure that none has been recorded within the last ten years. Now, the Saxons did know some of the regular summer migrants – the nightingale, the redstart and the red-backed shrike: all birds easily recognizable by some characteristic feature of song or habit – but it is really asking too much to expect them to be able to recognize a rare vagrant in winter. I think that we may be sure that the gyr-falcon and the eagle owl were not then rare vagrants. The possibility – the probability, as I believe – that in Saxon times both were resident and breeding in northern Britain cannot be dismissed.

But the great surprise is the vulture. This would have been the griffon vulture, which now nests no nearer than Spain. It has been recorded in Britain on only three occasions since 1843. But we do not know what was the breeding range of the griffon vulture in Saxon times. The bird is mentioned five times in Anglo-Saxon literature. The validity of the translation has been questioned by some authorities, but the possibility that the griffon vulture was a resident member of the British fauna in the seventh century cannot be dismissed out of hand. Of one thing we may be certain: if it was in Britain then, it would have found a good living. The Saxons were a bloodthirsty lot.

Chapter 5

1066 AND ALL THAT

THE YEAR 1066 is the one date in British history which everyone knows, no matter how unmindful of the past he may be. In that year William, Duke of Normandy, led the last effective invasion of this island. His invasion, like that of the Romans, but unlike those of the Saxons and the Danes (which were more in the nature of national migrations), was a deliberate military conquest carried out by a small army of professional soldiers, many of them mercenaries hired from all over Europe especially for the task. Unlike the Roman conquest, however, this was essentially an aristocratic conquest. William's small army was officered by the nobles of his court, each with his own company of retainers, fighting men in his permanent employ. A mere four years after the landing, in 1070 when the conquest was completed with the subjugation of the north, all the mercenaries were paid off and dismissed. William had won for himself a rich kingdom. But the regular army, so to speak, had still to be paid. This was done not by paying each individual soldier – that would have been wholly beyond his means – but by rewarding the nobles, his officers, with estates; the size of the estate depending upon the rank of the recipient. It was the business of the noble to satisfy his retainers.

William the Conqueror was one of the truly great men of history: a great king, a great soldier, and a superb administrator with the mind of a chartered accountant. He had, perforce, to parcel out his kingdom among his nobles. In doing so he was well aware that he was running a great risk, for these were powerful men with private armies at their command. So he took good care to know exactly what they were getting: and exactly what he could expect to get from them. Domesday Book, begun in 1085 and completed in 1087, was a marvellous enterprise, meticulously planned and admirably executed. It was a complete inventory of the nation, a valuation survey which told the Conqueror all that he needed to know: the owners of the land, the nature of its cultivation, the number of its inhabitants and their respective classes – freemen, villeins and serfs. It was at once a record of estate valuations and a census roll. It made possible taxation on a sound basis. And it left no loophole. From it he knew (and we know) not only exactly how many people there were in any one place at the time of its compilation, but also exactly how many pigs, how many sheep, how many cattle. With one exception wild life is ignored, for the very simple reason that it was not taxable and was, therefore, of no interest to the Conqueror.

The one exception is deer. William the Conqueror had a love of hunting which amounted to a passion. From Domesday Book we can form a fairly clear idea of the distribution of deer, especially in southern England. And the Forest Laws, which were chiefly concerned with the protection and conservation of deer, help us to fill in the picture to a considerable extent so far as some other animals are concerned.

'The forest', says Richard fitzNigel in the *Dialogus*, 'has its own laws, based not on the common law of the realm, but on arbitrary legislation by the king.' And he goes on to explain why this is so: 'In the forests are the secret places of the kings and their great delight. In them they go for hunting, having put off their cares, so that they may enjoy a little quiet. There, away from the continuous business and incessant turmoil of the court, they can for a little time breathe in the grace of natural liberty, wherefore it is that those who commit offences there lie under the royal displeasure alone.' The royal displeasure could not lightly be ignored. He that killed a deer died: he that shot at a deer – and missed – had his hands cut off: he that disturbed the deer had his eyes taken from him. The Conqueror took conservation seriously.

But he did not 'make' forests: he did not, for example, make the New Forest, as is commonly believed, All he did was to bring great areas of land under Forest Law. And he was careful in his choice of land. He never imposed his Forest Law on closely populated, well-cultivated land nor on the heavy clay lands. There was never, for example, a Royal forest in the whole county of Kent nor was any part of the Weald ever a Royal forest, because for the greater part of the year it was much too wet, so deep in mud as to be virtually impassable to horsemen. Nevertheless, huge tracts of England were made subject to Forest Law. Lady Stenton, in her *English Society in the Early Middle Ages*, points out that 'from the Thames, which for many miles was the northern boundary of Windsor forest, a man could in these days pass southwards, bearing in a somewhat westerly direction, through the woods of Eversley or Bagshot into Pamber and so into Bere Forest and finally into the king's New Forest which was bounded by Southampton Water and the Channel coast. Throughout his journey he would be within country which lay under forest law, and far on his either hand forest jurisdiction spread.' No one knows just how much of England lay within the borders of Royal forests. But it is known that almost the whole county of Essex was a Royal forest; that a great wide stretch of country, reaching from Oxford to Stamford, was subject to Forest Law; that most of the county of Cheshire and the whole of the Wirral peninsula was included within Delamere Forest. And there were also great tracts of country in Devon and Somerset, in Gloucestershire and Worcestershire, in Warwickshire and Shropshire, in Derbyshire and Nottinghamshire, in Lancashire and Yorkshire, in Cumberland and Northumberland, which were subject to Forest Law.

Not all these forests were on poor soil, of course: far from it. In areas where the population was small and thinly scattered huge stretches of good, fertile, thickly wooded country – England was then a very well-wooded land – lay within the boundaries of Royal forests. John Manwood, writing in the sixteenth century (but translating an early Norman work), described these forests as

> stored with great woods or coverts for the secret abode of wild beastes and also with fruitful pastures for their continual feed; for the want of either of these two, doth cause the exile of wild beastes from the Forest to some other place, for that the nature of the wild beaste of the Forest is, to flie into the thicke coverts for places of Secresie to rest in, whereof if there be none within the Forest, then they leave the Forest, and wander

> up and downe untill they find coverts elsewhere, which being without the bounds of the Forest, where these wild beastes are so found wandering, then they are hunted and killed to the utter destruction of the Forest: and in like manner it is, if the wild beastes have not those fruitful pastures within the Forest for their feed, then they pine away and starve, or else they are forced to forsake the Forest and to seek for food without the Forest, where they can find the same, and then they are likewise hunted and killed, whence it is manifest that a Forest cannot have continuance without wooddy grounds and fruitful pastures. And so consequently it followeth, that to destroy the coverts of the Forest, is to destroy the Forest itselfe: Also to convert the pasture grounds, meadowes, and feedings into arable land, is likewise to destroy the Forest.

A very clear exposition from the huntsman's, which was the Royal, point of view.

But William did not interfere with the people living within the borders of a Royal forest so long as they obeyed his Forest Law. He did not, as is so widely believed, destroy villages. He did not object to pigs being loosed in the woods at acorn time and he had no objection to land in the clearings being tilled, so long as it was land that was being tilled when he took over. But the acreage of such land within the boundaries of a Royal forest could not be increased. And, of course, the crops grown on it could not be protected against the ravages of deer. If a man wished to bring more land into cultivation, then he had to do so outside the boundaries of the forest.

This was part of a deliberate and far-sighted policy. William's Royal forests were not set aside solely for his personal enjoyment. There were far too many of them, covering far too large an area, for that. It is probable that in some he never hunted at all. That made no difference: all were thoroughly administered. The ultimate responsibility for administering a forest and for seeing to it that Forest Law was strictly observed lay with the chief forester, invariably a man of noble family: and a man who wielded immense power, for he was appointed by the king and was answerable only to him. The chief forester was supplied with a small office staff and a large field staff. Each forest was divided into sections under the direct control of a warden, who was appointed by the chief forester (though the post later tended to become hereditary) and who was responsible only to him. Below the wardens were the verderers, usually four to each county, who were elected by the shire court. They were unpaid and their chief work was to attend the forest courts – the attachment courts, as they were called – which dealt with offences against Forest Law and which were held every six weeks. The verderers were there to see that justice was done. But it was a foolhardy verderer who ran counter to the wishes of the chief forester. The ordinary day-to-day work of patrolling the forest was done by foresters, who were in effect gamekeepers. These men were appointed by the wardens and paid by him. Each was in charge of a comparatively small area, which he soon came to know intimately. Each knew full well that if he did not do his work properly he would certainly lose his job and his home; and possibly also his life. The Forest Law was rigidly enforced.

The system enabled the Conqueror to control development – in fact, to prevent it – within the boundaries of his Royal forests. More important; it enabled him to direct – in fact, to force – development outside them. He did not interfere with the hamlets and villages already established within the borders of the Royal forests nor with the land under cultivation around them. But neither could be increased. New settlement had to be made, new land taken up, outside the forest boundaries. Great areas of wild country lay outside these boundaries and these were not subject to Forest Law. In these areas a man might protect his crops: might hunt and slay the deer without fear of penalty. It was in these areas that additional cultivation or new building had to be undertaken. In areas where suitable land for cultivation was not readily available close to the forest boundaries men had either to move far afield or to tackle difficult land. In the comparatively thickly populated south of England, for example, there was, after the Conqueror's afforestation, so little suitable land available for cultivation, the good land having already been taken up by the Saxons, that some men were forced to make inroads into the Weald. But not many were prepared to undertake so arduous a task. Pressure on the easier lands steadily built up. A hundred years later Henry II greatly extended the boundaries of the Royal forests up and down the country. Henry II's passion for hunting was even greater than that of William the Conqueror and it is commonly believed that it was to satisfy his inordinate love of hunting that he enlarged the Royal forests. But that, surely, could not have been the sole reason. It is true that by bringing this land under Forest Law he afforded protection to the deer. But, at the same time, he stopped further development, and made farming more difficult, within the new boundaries. He was reinforcing the policy of William the Conqueror: forcing men to move, forcing them to bring new land into cultivation, relieving pressure on the easier lands. He, too, was a great and far-sighted king.

The animals protected by Forest Law were the red deer, the fallow deer, the roe deer, and the wild boar. These are precisely the animals that one would expect to find protected. What is surprising is that there is no mention of wild cattle. William fitz-Stephen, who died in the early years of the reign of Richard I, gives us a most valuable account of early London and the surrounding countryside in his life of Thomas à Becket. In it he mentions wild forest bulls, *Tauri sylvestres*, among the beasts living in the great woods which then reached to the very outskirts of London. It is quite evident that he did not mean the ordinary domestic bull; abundanly clear that he had some quite different animal in mind. Presumably, these wild cattle were the same as those mentioned in the laws of Hywel Dda; for it is surely inconceivable that there could have been two distinct types of wild cattle in Britain. But what were they? That they were not aurochs is certain. But, as I have said, the possibility that true aurochs' blood ran in their veins cannot be dismissed. It is not likely that aurochs bulls covered the cows of *Bos brachyceros*, the domestic cattle of Neolithic man (and a closely related species), from time to time? Is it not probable that cows of *Bos brachyceros*, which were not very thoroughly domesticated, attached themselves to the aurochs in the wild and so gave rise to a smaller race which yet had the genuine wild aurochs' blood coursing in its veins? There can, I think, be no other solution to the problem.

But this does not explain why these wild cattle were not protected by Forest Law. And it is surely surprising to find that they were not included among the animals which might be hunted by the king's subjects in his Royal forests. One can only suppose that, like the wolf, they were regarded as too dangerous to be afforded protection of any sort. The animals which might be hunted in Royal forests were those considered to be harmful to the deer: the wolf, the wild cat, the fox, the badger, and the squirrel (which was thought to interfere with the deer's feeding). In addition, permission was sometimes given to hunt the hare. (The protection of Forest Law was withdrawn from the roe deer early in the reign of Edward III because its numbers had increased so greatly that it was interfering with the other deer.) The right to hunt these animals was not given to the general public, but was granted by 'right of warren' to certain men only, and these were usually large landowners. They took good care, of course, to protect their rights, which made life even more complicated for ordinary folk living within the forest.

But nobody needed licence to kill a wolf. 'Let him bear the wolf's head' was the traditional form of sentence passed on a man declared an outlaw. It was the duty of the honest citizen to kill an outlaw on sight. It was no less his duty to kill a wolf on sight. And he who killed a wolf within the boundaries of a Royal forest could be sure of a reward from the king. In 1210 King John rewarded two men who killed two wolves at Gillingham in Dorset with fifteen shillings. This, in terms of today's values, was a very large sum indeed: and King John was not a generous man. That he was prepared to pay so large a reward suggests that he was not called upon to do so very often: is evidence that the wolf was already at that date a rare animal in southern England. And there is further, indirect evidence to support this contention. At this date none of the southern shires employed a professional wolf-hunter; presumably because there were not sufficient wolves to justify such an appointment. But all the shires along the Welsh border, some of the midland and all the northern shires, did employ professionals at this time. There is a specific reference to the numbers of wolves troubling country folk in the Yorkshire dales. The Worcestershire wolf-hunter at this date was paid the sum of three shillings per annum – compare that with the payment of fifteen shillings for the killing of two wolves – and there is no evidence that he was called upon to account for the number of wolves he killed. It would be only when the authorities began to wonder whether there were sufficient wolves to justify the employment of a full-time professional that he would be called upon to account for the number killed. It would seem incontrovertible that at this time the wolf was still plentiful along the Welsh border and in the north.

Henry II was the last monarch to enlarge the Royal forests. He was the last monarch rich enough and powerful enough to be able to do so. Administering the forests, enforcing Forest Law, was a difficult and expensive business, which became more difficult and more expensive with each passing year, for Forest Law was bitterly resented not only by the humble folk within the forest but also by the nobles and the Knights of the Shire (the country gentlemen) and the freemen who lived outside. This was a land-hungry nation. Those within the forests resented being forced out, often on to less suitable land, to better themselves, when they saw so much good land all around them which could not be touched: those outside were equally resentful, for the movement constituted a danger to their own holdings. William the Conqueror, an absolute monarch ruling a defeated and cowed country, had no difficulty in imposing his will. But defeated countries do not remain cowed for ever: and gratitude is not an emotion noted for endurance. The nobles rewarded for services rendered were grateful: they also knew that quarrelling with the Conqueror was a pursuit with no future. The humble within the borders of a forest, cowed by the Conqueror, accepted the Forest Law, for they dared not do otherwise: retribution was too swift and too terrible. But this situation could not last. With the death of the Conqueror the nobles forgot to whom they owed their lands: they remembered that they owned them, that they were rich and powerful, and that they would like more. And spirit returned to the humble. To a meat-hungry people the Royal forests were full of temptations. They began to poach on a very considerable scale. To kill a deer within the borders of a Royal forest was altogether too risky. But it was perfectly legal to kill deer beyond the boundary. It became the custom, wherever possible, to drive deer over the forest boundary, so that they might be killed without fear of punishment. Of course, those caught driving deer paid the penalty: blinding or maiming. But they were none too easy to catch. To prevent deer being driven necessitated the employment of a larger number of foresters, and the larger the forest the larger the force of these men needed. Henry II was the last monarch able to afford such a strain upon his purse.

With his death the process of disafforestation commenced. Richard I, who spent practically the whole of his reign on expensive adventures abroad, was always short of money. In the very first year of his reign the Knights of Surrey, the country gentlemen of Surrey, knowing well with whom they had to deal, offered him two hundred marks – a huge sum in those days – 'that they might be quit of all things that belong to the forest from the water of Wey to Kent and from the street of Guildford southwards as far as Surrey stretches'. The offer was accepted. A large area of Surrey was thus freed from Forest Law, protection was withdrawn from the deer, and the land was gradually brought into cultivation. Large-scale disafforestation on similar lines also occurred in several areas of the midlands. Furthermore, Richard I was prepared to allow 'assarts' in return for hard cash. Richard fitzNigel defines 'assarts' as 'cutting down the forest, woods, and thickets suitable for feeding animals, ploughing the soil and cultivating it'. Give a countryman an inch and he will take an ell. The monarch was abroad, nobody bothered very much. A large number of unofficial 'assarts' were made in the huge area of Royal forest up and down the country. Towards the end of his reign, short of money as usual, Richard I caused a searching survey to be made of these unofficial 'assarts'. Many were caught in the net and made to pay half a mark each; a considerable sum in those days. In every case the fine was paid in full. But the important point is not that the fines were paid, but that the offenders, having paid, were left in peace to enjoy the land they had filched in return for a rent. It meant that the King accepted that encroachment upon the Royal forests was inevitable. King John also accepted that fact. King John had three passions in life: hunting, money, and women. And the greatest of these was money: during his reign he acquired a vast treasure. Naturally, the cost of disafforestation rose sharply. The gentlemen of Essex had to pay 500 marks for the disafforestation of that part of Essex which lay north of the Bishop's Stortford–Colchester causeway; the gentlemen of Cornwall 2,200 marks for the disafforestation of the whole county; the gentlemen of Devon no less than 5,000 marks (an enormous sum) for the disafforestation of the whole of their county. These were extortionate prices. But men wanted land and always the money was found. The cost of 'assarts' also rose abruptly; and again the money was always found. But King John, though he was always willing to sell parts of the Royal forests, was also passionately fond of hunting and, where he did not sell, he enforced the Forest Law with regard to deer with the utmost rigour. That side of the picture changed under Henry III. He was not interested in hunting. His interest was not in venison, but in vert: in the greenwood of the forest. In his reign a much more serious view began to be taken of offences against the vert: a man was likely to get a longer term of imprisonment for felling a grown tree than for killing a deer. (Henry III abolished the death penalty for killing a deer in a Royal forest.) So keenly interested was he in the vert that he even re-afforested land which had been disafforested, thereby making himself extremely unpopular and ensuring that no future monarch – not even so powerful and well-loved a king as his son Edward I (who tried) – would ever be allowed to re-afforest land.

The important point about this – and from the point of view of wild life in Britain for the next five hundred years it is very important indeed – is that the land bought (or filched) from the Royal forests became subject to very much the same restrictions. The ownership of the land changed, but not, basically, the policy regarding game. The Royal lesson had been well learned. Forest Law disappeared from these areas, but its place was taken by, so to speak, private law, which was not generally very different and which was just as rigorously enforced. This did not meet with the approval of royalty, but there was little they could do about it. In 1389 Richard II accepted the position, passing an Act which allowed anyone qualified by estate – and the qualification, considering the times, was quite remarkably liberal – to kill game.

The principle embodied in this Act remained in force until the Game Act of 1831.

Protection was withdrawn from the deer with disafforestation. But the new owner, though he took good care to protect his crops, also took good care to protect his deer. He saw to it that their numbers did not grow too great, but he also saw to it that unauthorized persons did not help themselves to his winter meat supply. He fenced his deer in. The custom of making parks or enclosures from which deer could not stray, and from which they could not easily be driven, spread all over the country. Many of these parks also included a fish-pond, usually formed by damming a stream, so that the landowner might get his fast-day food without difficulty. This led, inevitably, to an increase in poaching, both of deer and fish. Guards (not yet gamekeepers) began to be employed and some of the larger landowners had their private prisons into which poachers were thrown without trial.

The introduction of the rabbit to England – an introduction which, in due course, was to have an enormous effect on wild life generally – must surely date from the establishment of the private, fenced-in park, for it would have been introduced for its meat and the introducer, not a philanthropist, would have hoped to contain it. The rabbit is not mentioned in Domesday Book nor is it included in 'right of warren', which was first granted by Henry I, nor in any title to 'right of warren' granted by Henry II. Its introduction almost certainly took place early in the thirteenth century: perhaps towards the end of the reign of King John, more probably during the first half of the reign of Henry III. The first absolutely certain record of the rabbit in England is for 1272. But it must have been introduced a good deal earlier than that, for two thousand skins were taken from Lundy Island in 1274. Since rabbits fetched a high price throughout the thirteenth and fourteenth centuries, it is evident that they were not then widespread – and, indeed, they do not appear in the statute book until 1389 – which suggests that Lundy Island may have been the original place of introduction. If this was so, then Sir William de Marisco or his son, Sir Otho, lords of the island, would have been responsible. Both are known to have had connections with Spain, where the rabbit was common. During the fifteenth century rabbit formed part of the menu at great banquets – no less than four thousand were provided for the banquet following the installation of George Nevill as Archbishop of York in 1465 – which is evidence that they must by then have become fairly widespread, but were still sufficiently uncommon to be regarded as a luxury. In the sixteenth century they were certainly widespread and fairly abundant – sufficiently so to have disappeared from the menu at state banquets – but they do not seem to have been really common. They are mentioned ten times by Shakespeare (sometimes as rabbits, sometimes as coneys) and one of these references is certainly to poaching – Shakespeare, if legend is to be believed, was himself a bit of a poacher in his youth, though with a preference for the deer in the local squire's park – but all are passing references. One has the feeling that Shakespeare knew that the rabbit would not be really familiar to the majority of the audiences it was his business to entertain. Though by the middle of the seventeenth century the rabbit warren had become a feature of many estates – though rabbits were then virtually farmed for their meat and fur – it was not until the change in agricultural practice brought about by the Enclosures that the animal became really common.

The black rat, an accidental introduction which was to have an enormous effect on British history, was in England before the rabbit. There seems to be general agreement that it reached this country on ships returning from the Crusades. This does not, however, give us a date. The Crusades, the Wars of the Cross, covered a period of over two hundred years; from about the middle of the eleventh century until the end of the thirteenth. Scarcely a decade in all that time passed without one or more expeditions, most of which – and certainly all the big ones (the ones known by numbers) – included Englishmen. The Third Crusade, that associated with Richard I, is popularly favoured. But the black rat was certainly in Britain, and in large numbers, long before then. Perhaps it would be safe to place its introduction at some time during the first half of the twelfth century.

A few people – on the evidence of Archbishop's Aelfric's Anglo-Saxon Vocabulary, which was compiled about A.D. 1000 – would put it at long before that. But there is some doubt as to whether 'rat' is the correct translation of *raet*. Some scholars, pointing out that *raeta* was the Provençal word for house mouse at the time, believe that it is not. Even if one accepts the translation as correct, that does not necessarily mean that the rat was present in Britain in A.D. 1000. It is surely significant that there is no other mention of it whatsoever. And it is impossible to believe that so formidable an animal, and one so closely associated with man, could have escaped notice by the very observant compilers of Domesday Book.

The earliest undoubted reference to the rat in Britain is to be found in the work of Giraldus de Barri (1146–1220), commonly known as Giraldus Cambrensis, who was Archdeacon of Brecon. He was the first man clearly to differentiate between rat and mouse. He does so in the *Itinerary Through Wales*, which was written towards the close of the twelfth century. And he does so unmistakably: *murium majorum, qui vulgariter Rati dicunter*, 'the larger kind of mice, commonly called Rats'. Giraldus is telling the story of a man being devoured by rats. There is no reason to doubt the story – there are several similar records of much more recent date, including one for 1903 in the United States (brown rats being responsible in this case) – and it provides some indirect evidence that the rat was well established, was present in considerable numbers, in that region at least at that time. From the time of Giraldus onwards mention of the rat in Britain becomes more and more frequent. Before the end of the thirteenth century it was a nation-wide pest and was greatly feared because its connection with disease was at least vaguely appreciated. It was the black rat which was responsible for the Black Death of 1348, the outbreak of plague which killed between one-third and one-half of the entire population, bringing about far-reaching changes in the relations between landlord and peasant and altering the whole aspect of English rural life. It was the black rat which was responsible, some three hundred years later, for the Great Plague of London. But it was not only its connection with terrible epidemic diseases which made the black rat feared and hated: it was also the enormous damage it caused day by day, in town and country alike, to the works of man. In Shakespeare's time days of prayer for protection against its ravages were set aside and the rat-catcher was a very important official indeed. Shakespeare makes many references to the rat – he had no doubt about his audiences' familiarity with the animal in this case! – and one definite reference to the rat-catcher.

To these two introductions – the one deliberate, the other accidental – must be added a third. The pheasant, which like the rabbit was in due course to have an enormous effect upon British wild life, was deliberately reintroduced during this period, probably during the reign of Henry III, but even the approximate date is not known. And it was during this period that one mammal disappeared from the British fauna, exterminated by man. This was the beaver.

Giraldus Cambrensis tells us that on his journey through Wales with the Archbishop of Canterbury in 1188 he found beavers on the River Teifi. There is no doubt that he actually saw them. He gives a long account of the methods used by beavers in building their dams and of their habits and anatomy, and this could all have been obtained at second-hand, but he also describes their appearance and the use of their tails; and this has the stamp of the eye-witness about it. Giraldus tells us that, at that time, beavers were to be found nowhere else in England or Wales. In England beavers had, indeed, been exterminated at a much earlier date, for there is no reference to them in Anglo-Saxon literature nor in Domesday Book. Had they been present in England at the time of that great survey there can be no doubt that their presence would have been duly recorded, for the fur was highly prized by the nobility and fetched roughly ten times the price of any other animal skin. But when did the beaver become extinct in Wales? In Holinshed's *Chronicle* (1577) we find 'to saie the truth we have not manie Bevers but onelie in the Teifie in Wales'. But that bears the stamp of Giraldus, not of personal observation. John Leland, who made a journey through Wales about 1536, recording whatever he saw of interest (and he was interested in natural history as well as in antiquities), spent some days in the Teifi area, but makes no mention of the beaver. I think there can be little doubt that the beaver was extinct in Wales before the beginning of the sixteenth century. It lingered but little longer, if longer at all, in Scotland. Hector Boece, the chronicler of Scottish history, writing in 1526, mentions beavers being trapped in Loch Ness and says that their fur fetched high prices in foreign markets. But it is not clear whether he is writing of the present or of the recent past. This is the last record of the beaver in Scotland that I have been able to find. It may be taken, I think, that that particular export trade ceased at about that time.

During the civil wars of the fifteenth century, the Wars of the Roses, the administration of the Royal forests completely broke down. It is important to realize that these wars were really quarrels between great families struggling for supremacy rather than wars of ideals. The mass of people seem to have been quite extraordinarily unaffected. But the old nobility did virtually wipe themselves out. Their place was taken by a new class of property owners, a merchant class, with smaller fortunes: a class without estates, a class which required land and houses. Added to the steadily growing pressure on land, there was now an ever-increasing demand for building material. As administration of the Royal forests had broken down, men helped themselves: the peasants to the underwood for their cottages (which were little more than hovels of wattle and daub), the merchant class to timber for their houses. In the midlands particularly there was a great deal of timber felling. This led to an important discovery – that, once cut, the deer and the grazing animals of the local populace never allowed the underwood or seedling trees to grow again – which resulted in the first law in English history affecting silviculture being enacted in 1483. By this Act enclosure for a period of seven years was made legal. Where underwood and timber had been cut out, it permitted a ditch to be dug and a bank thrown up and planted with thorns so that the area might be protected from the assault of grazing animals. At the end of seven years, when these animals could no longer do serious harm, the land had to be thrown open again. It was still, of course, illegal to cut underwood or fell timber in a Royal forest without permission and payment, but the Act is evidence that the position had got completely out of hand, that the law could no longer be rigorously enforced. The Act accepted the fact and tried to make the best of it.

What actually happened was that men enclosed areas within a

forest, as they were legally entitled to do: but they often failed to throw them open again at the end of seven years. If careful record was not kept, regular inspection not made, such enclosures tended to be absorbed into private ownership, to be settled and built upon. And, of course, once settled and the fine (if caught!) paid, the place became an estate under the 1389 Act. This did not happen in the New Forest – a survey made by Elizabeth I showed that 5,800 acres had been enclosed without permission or payment, but the timber was used for building outside the forest: the soil was too poor to be worth settling – but it did happen all over the midlands. The great midland forests – Arden, Delamere, Rockingham – began to disintegrate during the Wars of the Roses. By the end of the sixteenth century much of England had assumed the aspect it was to retain for the next two hundred years and more.

It was during the Wars of the Roses, just about in the middle of them, in 1460 or thereabouts, that the great bustard was first recorded in England. It may be taken as beyond question that it was not in Britain before this date. So large a bird – the great bustard stands about three feet six inches, weighs around thirty pounds, has a wing-spread of seven to eight feet, and is by far the largest bird ever recorded as resident in Britain – could not possibly have been overlooked, for the simple reason that it offered a sizeable meal and the English countryman has never been one to ignore such an offer. Indeed, all the early references to the great bustard are to its use for the table. It is tempting to think that it was introduced, as was the pheasant, by some magnate as a decoration for his park and to provide an addition to his table on special occasions: and I suppose, even though the times were hardly propitious for such an experiment, that that is just a possibility. But it is much more probable that the great bustard's arrival in this country was the result of a natural extension of its range; as has recently happened in the case of the collar dove.

Of the birds of the period we know very little. According to James Fisher only about one hundred and fifty species were listed at the end of the sixteenth century. There were, of course, many more: and certainly more breeding species than there are today. It was ornithologists, not birds, that were scarce. The numbers of some of these birds would have been very different from today and particularly would this have been true of the birds of the wet lands of eastern England.

The Romans had made a good job of draining a considerable area of the Fens. By the time of the Norman Conquest their work had, of course, fallen into decay. An unsuccessful attempt was made to drain Deeping Fen in the reign of William the Conqueror. After this the whole district was abandoned: only isolated spots rising above the surrounding level being occupied by monasteries. These did a certain amount of drainage in their immediate neighbourhood, but only sufficient to provide cultivation for themselves. And even that little stopped abruptly with the dissolution of the monasteries in the reign of Henry VIII. No further attempt at drainage on a large scale was made until the reign of Charles I. So, for the whole of this period, the area must have looked very much as described in the earliest English poem, which celebrates the deeds of Beowulf: a land of mere, mist, and morass, a land of evil spirits and the home of many thousands of wildfowl. We cannot even guess at how many thousands. But it is known that four thousand mallards were brought to table at the banquet for George Nevill in 1465 and that all came from Deeping Fen. William Camden, in his *Britannia* (1607), says that three thousand mallards were taken 'at one draught of the net' in Lincolnshire in August, but he does not, unfortunately, name the place where this haul was made.

Among the birds nesting in the Fens in the sixteenth century were the crane and the spoonbill – the latter was also nesting in 1524 in the park of the Bishop of London at Fulham – but both must already have been becoming scarce because an Act of 1534 granted them and their eggs protection between 31st May and 1st August. This Act also gave protection for that period to the bittern, the great bustard and the heron, and to their eggs.

Protection was also afforded the kite: but for a different reason. As the towns grew, so did the refuse. The kite was greatly valued as a scavenger at a time when there was no refuse collection. Schaschek, an Austrian who spent some time in London in 1461, recorded that he had never seen so many kites as there were about London Bridge and was much impressed by their graceful flight. Belon, a French doctor who visited London in the middle of the sixteenth century, thought the kites to be as numerous as in Cairo. William Turner, the father of British ornithology, writing about 1540, says of the kite: 'This kind is wont to snatch the food out of the children's hands, in our cities and towns.' Sufficient indication of tameness or rather of boldness born of immunity.

Chapter 6

THE FIRST GUN AND THE LAST WOLF

As a breeding species, the crane became extinct in Britain at the end of the sixteenth century. Nobody knows the exact date, but it was probably not later than 1600. Nor is it known why it ceased to breed in Britain. But it is certain that man was not to blame. There was no drastic interference with the wet lands of eastern England at that time. Moreover, it continued to visit eastern England, and in large numbers, for many years after it had ceased to breed there. Willoughby, writing in 1676, says: 'They come to us often in England, and in the fen counties, in Lincolnshire and Cambridgeshire, there are great flocks of them.' Some birds do, from time to time, and for no apparent reason, change their habits. The crane would appear to be an example of this.

As a breeding species the spoonbill lasted a little longer. Again, nobody knows the exact date when they ceased to breed in Britain, but Sir Thomas Browne records them as breeding in Trimley in Suffolk about 1667. There does not appear to be a later record, and it would probably be safe to say that the spoonbill was extinct as a breeding species well before the end of the seventeenth century. In this case – as in many others concerning birds of the wet lands – man must certainly bear the blame; or a great part of it.

In 1630 the first really determined effort since the Romans was made to drain the Fens. Cornelius Vermuyden, a Dutch engineer who had already successfully drained 70,000 acres of marshland in Yorkshire, was appointed to carry out the Earl of Bedford's project to drain the Great Level of the Fens. The work took seven years to complete, but further work went on under the Bedford Level Corporation, with Vermuyden as its chief engineer, until 1655. Indeed, drainage work in the Fens never really stopped, for further great works were undertaken in the eighteenth and nineteenth centuries. However, it is with the work between 1630 and 1655 that we are concerned here. The Fenmen did not like Vermuyden and they did not like his work. It interfered with many of their 'rights' and there were many complaints and not a little violence. The 'rights' complaints were soon dropped – it would seem that many Fenmen held land to which they had no title in law – and by 1653 the general complaint was that the Fens were 'spoiled for fish and fowl'. Unquestionably, there was truth in this. The fish must have been most gravely disturbed and certainly the fowl suffered greatly.

We have seen what happened to the spoonbill. But the spoonbill was only one species among many. Before Vermuyden commenced his great drainage work the avocet, the black-tailed godwit, the bittern, the ruff, the marsh harrier, and Savi's warbler – and perhaps also other warblers: James Fisher has suggested both the aquatic and the melodious – were breeding regularly in the Fens. By the time Vermuyden's work was completed all these were greatly reduced in numbers or had wholly disappeared as breeding species from the Fens; and by the completion of the nineteenth-century schemes as breeding species from Britain as a whole. (A few have since returned, in small numbers, and these will be considered later.) Now it is important to understand that none of these species was deliberately exterminated. Such a thought never entered the head of Vermuyden: never entered the head of any one of the later engineers. They simply were not interested in birds. The attack was upon the Fens, not upon the Fenland birds. The birds disappeared because they were subjected to severe, if quite unconscious, harassment during the breeding season and because their natural habitat was so savagely interfered with as to be virtually destroyed. Many birds have proved themselves to be highly adaptable in their relation to man and his works. Marsh birds, on the other hand, have, time and again, proved themselves unable to adapt to new conditions.

By the end of the sixteenth century the reshuffling of the aristocracy consequent upon the Wars of the Roses had become an accomplished fact and the 'new nobility' was indistinguishable from the old: except, perhaps, in the eyes of the old! In the seventeenth century the 'country gentleman' class, which was to play so important a part in the rural history of Britain in the next three hundred years, began to develop as an offshoot of the merchant class. These country gentlemen had an instinct for profit. They began to pay attention to improved methods of agriculture – for example, they played a large part in the employment of Vermuyden and the drainage of the Fens – but they also began to pay attention to country pursuits which had hitherto been regarded as largely the prerogative of royalty and the aristocracy, the chief of which was hunting. And the time was ripe for them. The Tudors, on the whole, were much too busy to have time for hunting. There was a gap to be filled and they filled it.

In Tudor times there were still animals to be hunted in Britain which have long since disappeared: notably, wild cattle and wild boar. George Owen recorded how he watched wild cattle hunted on the Pembrokeshire hills in the reign of Elizabeth I:

> I have also seene good pastime in hunteing and killing the wild bull, wild oxe and wild calfe, by horsemen and footemen, whereof there is yett some store, reared upon the mountaines, thoughe lesse than heretofore, the owners finfinge more profite by the tame, than pleasure in the wilde.

These would, presumably, have been the same as those recorded by fitzStephen as being present on the outskirts of London some four centuries earlier: again the wildness is stressed, but there is no mention of great size. And there were still wild cattle on the mountains of Wales half a century later: that is, in the early years of the seventeenth century. In the *Report of the Royal Commission on Land in Wales and Monmouthshire* (1894–6) there is an extract from a letter written by an Englishman:

> Wild calves are common in Wales upon the mountains, whence one was brought this last Christmas to Ludlow Castle, when I did eat of it, roasted and bak'd: and by taste I find it more firm and dry, and by the efforts of digestion more wholesome and passable than our ordinary veal.

'Common' may, I think, be questioned. We know that their numbers were declining in Pembrokeshire fifty years earlier and this is the last report we have of wild cattle in Wales. In any case, they could not, in the very nature of man and his affairs, have survived much beyond this time. But of 'wild' there can surely be no doubt. This man was being given something unusual – a Christmas treat – and he knew it. These were wild cattle: not domestic cattle put out on the mountains to graze, as are the Welsh Blacks on Plynlimon today. No man in his senses would then have put domestic cattle out on the mountains, for there were still wolves in Wales. These wild cattle would not have been exterminated by hunting. Since there was 'more profite by the tame', they would have gradually been absorbed into the domestic herds.

The wild boar disappeared in the same manner. It just faded out. It was not hunted to extinction: it was brought to extinction by a combination of hunting and domestication. Semi-domesticated pigs – pigs which, though smaller and lighter in body, resembled the wild boar in general appearance, being long in the leg, having very long snouts (much longer than those to be found in any modern breed of domestic pig), and having a ruff of coarse bristles round the neck and a pronounced ridge of coarse bristles along the spine – were common in the Scottish Highlands until the latter part of the nineteenth century. These pigs foraged for themselves on the moors and sometimes caused considerable commotion by raiding corn crops. For the greater part of the year they led what was virtually a wild life, but all had owners (at least in theory) and most were brought in for the worst of the winter. Rounding them up could not have been easy, for they were gifted with quite remarkable speed. Pigs, similar in appearance to these Scottish pigs, were also fairly common on the Welsh mountains up to the middle of the nineteenth century and they, too, foraged for themselves for the greater part of the year. I think that there can be little doubt that these 'razorbacks' had wild-boar blood in their veins. When pig-breeding began to be taken seriously and improved breeds became available from England they rapidly disappeared, first in Wales and not very long afterwards in Scotland. And with them went the last trace of the wild boar.

But there were still wild boars to be hunted in Elizabethan England and in Jacobean England also. James I was probably the last monarch to hunt the wild boar. He was a keen hunting man and on his way south for his coronation, having stopped a night at Newark, he hunted all the way down to Belvoir Castle. The quarry is said to have been wild boar. But the wild boar was certainly extinct in the midlands and the south of England by the reign of Charles I, for he is known to have imported a number from France and released them in the New Forest where, John Aubrey says, they 'became terrible to the travellers'. They were all destroyed during the Civil War. But there were still wild boar in the north of England a great deal later than this, for they are mentioned in an account book of the Manor of Chartley in 1683 and they may well have lingered on in the wilds of the north until the early years of the eighteenth century. Another attempt at re-introduction in the south of England was made by General Howe about 1702. He imported a number from Germany and released them in Wolmer Forest. 'But', says Gilbert White, 'the country rose upon them and destroyed them'.

While on the subject of extinction we may as well get the wolf out of the way. In England as a whole the wolf became extinct early in the sixteenth century: the exact date is not known. But it was to be a further two hundred years and more before the last wolf in Britain was killed. Gradually, over the centuries, their range was restricted to the wilder and more northerly regions, and there their numbers were greatly reduced not so much by actual hunting as by burning the woods. A monument by the roadside at the foot of Glen Loth commemorates the slaying of the last wolf in Scotland by a man named Polson about the year 1700. The monument appears to be of late Victorian date and the meagre information, the doubt about the date, is unconvincing to say the least. It is, in fact, inaccurate. Wolves were present in Scotland at a much later date.

It is generally accepted that the last wolf in Scotland was killed by Macqueen of Pall-a-chrocain in Tarnaway Forest in Sutherland in 1743. The story is told by J. S. and C. E. Stuart in *Lays of the Deer Forest*, and is worth repeating.

> One winter's day Macqueen received a message from the Laird of Mackintosh that a large 'black beast' supposed to be

a wolf had appeared in the glens, and the day before killed two children who with their mother were crossing the hills from Calder; in consequence of which a 'Tainchel', or gathering to drive the country, was called to meet at a tryst above Fi-Giuthas, where Macqueen was invited to attend with his dogs. Pall-a-chrocain informed himself of the place where the children had been killed, the last tracks of the wolf, and the conjectures of his haunts, and promised his assistance.

In the morning the 'Tainchel' had long assembled, and Mackintosh waited with impatience, but Macqueen did not arrive. His dogs and himself were, however, auxiliaries too important to be left behind, and they continued to wait until the best of a hunter's morning was gone, when at last he appeared and Mackintosh received him with an irritable expression of disappointment.

'Ciod e a'chabhag?' ('What was the hurry?'), said Pall-a-chrocain.

Mackintosh gave an indignant retort, and all present made some impatient reply.

Macqueen lifted his plaid and drew back the bloody head of the wolf from under his arm. 'Sin e dhuibh' ('There it is for you'), said he, and tossed it on the grass in the midst of the surprised circle.

Mackintosh expressed great joy and admiration and gave him the land called Sean-achan for meat to his dogs.

The Stuarts had that story in 1830 from an eighty-year-old man, whose father had been present at the 'Tainchel' and it bears the ring of truth: the dramatic gesture so beloved of the Highlander, the sudden generosity, the quixotic sentimental generosity, so typical of the Highland laird. This (though the story will not, of course, have lost in the retelling) is what happened.

But Macqueen's wolf, though the last in Scotland, may not have been the last in Britain. There is just a possibility that the wolf may have lingered on in Wales for a further forty years. Colin Matheson, in his *Changes in the Fauna of Wales within Historic Times*, mentions that Robert Prys Morris, in his *Cantref Meirionydd*, refers to the reported killing of a wolf near Dolgellau, at a place known as Ogo'r Fleiddiast (Wolf's Cave), in 1785. Matheson dismisses this report because the story as recounted by Morris is very similar in detail to the story told of the killing of one of the Scottish wolves and also because of the lateness of the date. I think that the killing of one wolf is very likely to have resembled, even in points of detail, the killing of another. But the date is certainly very late indeed. One would have expected to find records of the presence of wolves during, say, the previous forty years. So far as I am aware there are none. I think that Matheson is absolutely right to dismiss this report. But the possibility, no matter how remote, must be mentioned.

The killing of wolves was anybody's business and nobody bothered about the method used. Hunting, as such, was the relaxation of the gentry. In Tudor times, and at least until the middle of the eighteenth century, the hare was the chief quarry. The fox was generally regarded as vermin: to be destroyed by any means. The most usual method at this time was that which had been used for many generations past: namely, to net the sides of a copse and to rattle the fox about with dogs of any kind until it ran into the net, when it would be despatched with sticks. This may sound to us a rather haphazard method of dealing with so clever an animal as the fox, but it must have been pretty effective, for we find the chronicler Holinshed writing to Queen Elizabeth to say that if foxes were not preserved for the pleasure of gentlemen they would be entirely exterminated. 'So earnestly are the inhabitants bent to root him out', he wrote, 'that except it be to bear with the recreations of their superiors they must have been utterly destroyed manie years agone.' This suggests that there was some fox-hunting at the time. Yet there would seem to be only one record of a fox-hunt in Elizabeth's reign. That was near Eastbourne and the dogs used were greyhounds. The fox seems to have escaped. It was the hare that interested the gentry. They still subscribed to the view of Edward, Duke of York who, in his book *The Master of Game*, written in the fourteenth century, says: 'It is to be known that the hare is king of all venery, for all blowing and the fair terms of hunting cometh of the seeking and the finding of the hare. For certain it is the most marvellous beast that is.' And he has this to say on the proper hunting of the hare:

'It is a fair thing to slay her with strength of hounds, for she runneth strong and cunningly. A hare shall last well four miles or more or less, if she be an old male hare'. Shakespeare, who knew about everything, knew also about hare-hunting.

> My hounds are bred out of the Spartan kind,
> So flew'd, so sanded; and their heads are hung
> With ears that sweep away the morning dew;
> Crook-knee'd, and dew-lapped like Thessalian bulls;
> Slow in pursuit, but match'd in mouth like bells,
> Each under each. A cry more tuneable
> Was never holla'd to, nor cheered with horn,
> In Crete, in Sparta, nor in Thessaly;
> Judge when you hear.

The basset hound, I know, was not introduced to England until 1866, but that is as good a description of basset hounds as I have read anywhere: and of their work, too. There can be no doubt that Shakespeare knew about hare-hunting from personal experience: that he appreciated the work and the music of hounds, and he was more interested in the hunt than in the kill.

'Slow in pursuit.' The work of the hounds was very slow judged by modern standards. But, then, so was life. There was not so much to do, and there was a great deal more time in which to do it. Our forbears went about life in a much more deliberate manner. They used to start hunting at a very early hour, and they used to start by hunting the trace the hare had left during her feeding the previous night, unravelling it as they went along, until finally they put her up from her form. Then the hunt proper began, and frequently it was a pretty long one. In *Of the Hunting of an Hare* it is impressed upon the huntsman that he should mark carefully the first runs and circles of the hare, for 'she will all day long hold the same wayes'. One hare; all day long. And, indeed, there is an account of a day with Sir Roger de Coverley's pack, at the end of which the hare is said to have been picked up alive eight yards in front of the hounds, who had been hunting her for nearly as many hours. 'Slow in pursuit': hare-hunting was a leisurely sport matching a leisurely life. And it was the principal sport of the majority of country squires and farmers during the sixteenth and seventeenth centuries. Probably most country squires and some yeoman farmers kept a few couple of hounds with which they hunted whenever they felt inclined. Indeed, it seems likely that there are very few villages in England and Wales that have not had at some time their pack of hounds. This sort of hunting, no matter how many packs of hounds were in operation up and down the country at the same time, did not affect the balance of wild life at all.

It was the gun that made the difference. It was the gun that was to alter the whole complexion of country life. Guns had been fired at game – I use the term comprehensively – on the Continent before the end of the fifteenth century. The wheel-lock was invented at the beginning of the sixteenth century, and a primitive breech-loader was in use as early as 1537. It is known that wildfowl were killed with guns in Norfolk in 1533, but the crossbow was then, of course, still the principal weapon. It is often said that Roger Ascham, who was Latin secretary to Queen Mary and who became private tutor to Queen Elizabeth, wrote the first book on shooting in the English language. This, as he makes plain, in his introduction, is true enough:

> That no man hitherto hath written any booke of shootinge, the faulte is not to be layed in the thinge which was worthye to be written uppon but of men which were negligente in doinge it, and this was the cause thereof as I suppose. Menne that used shootinge most and knewe it best, were not learned; men that were learned, used little shootinge, and were ignorant in the nature of the thinge, and so fewe men have been that thitherto were able to write uppon it.

But his *Toxophilus*, published in 1545, was a treatise on archery: evidence enough that the gun was not taken seriously as a game weapon at that time. It is also evident from Ascham's work – indeed, it is implicit in the whole of it – that birds, and particularly wildfowl, were then regarded as the proper quarry for the bow.

> Yet welfare the gentle goose whiche bringeth to a man even to his doore so many exceeding commodities. For the goose is

> man's comfort in warre and in peace slepinge and wakinge. That praise soever is given to shootinge, the goose may challenge the best part of it. Howe well doth she make a man fare at his table! How easelye doth shee make a man lye in his bedde! How fitte even as her fethers be onlye for shootinge, so be her quilles fitte onlye for wryting.

Men hunted the deer and the hare, trapped the fox or chivvied him with mongrels, but they shot birds. It was natural, therefore, that, when the gun replaced the bow, birds should remain the chief quarry of the shooting-man.

It took a long time for the gun to replace the bow. In 1586 a cartridge was devised for the muzzle-loaders of those days – the first breech-loaders do not seem to have had any success and were widely regarded as most dangerous to the men who used them – and gun manufacture, on a small scale, commenced in Birmingham in 1603. Paris was then the centre of the craft, but very few French guns seem to have been imported. English guns were pretty primitive and most men still preferred the crossbow as being the more efficient weapon, but sufficient guns were in use to prompt Gervase Markham to write the first book in English on game-shooting as such. His *Hunger's Prevention or The Whole Arte of Fowling* was published in 1621. In it he had this to say of the gun of the time:

> The next Engine to these is the Gun or Fowling-Piece, which is a generall Engine and may serve for any Fowle great or little whatsoever, for it hath no respect at what it striketh, being within the levell: And of the Fowling-Piece you shall understand that to be the best which is of the longest barrell as five foote and a half, or sixe foote, and the boare indifferent, as some what under Harquebush, for these hould the best charges and carry the furthest levell, which is a principal thing to be regarded; for the Fowle are of such a tickell and cunning nature, that a man shall hardly get within any indifferent or neare station, and to shoote out of levell were to shoote against the winde, and scare-Crow like, onely to affright Fowle with loss of labour.
>
> Now for the use of this Fowling-Piece, it is either for Land

> or Water that is passable or may be waded, as whether it bee in Fenne, upon Plash, or Pits, or any other blank water; and herein first the Fowler is to observe the finding out of his Game, and which Fowle lyeth fittest for his purpose, at no time striving to shoote at a single Fowle, if he can by any means compasse more within his levell; then he shall seeke as neare as he can to shoote with the winde, not against the winde, and rather side waies or behind the Fowle than full in their faces; then he shall seeke the convenientest shelter he can finde as either hedge, banke, tree, or any other shaddow which may hide him from the gaze of the Fowle, for they are so fearfull of the proportion, visage, or motion of a man, that upon the least suspicion they are gone in a moment, and therefore when under his covert or pertition he is gotten within his levell and hath the winde fit and certaine, then he shall make choice of his marke, which (as before I said) would in no wise be a single Fowle, but rather the longest and largest Rancke or File of Fowle you can finde, for since one shoote is as much as you can get at one time and in one place, it is meete you put it to as great use and profit as you can possibly...

The odds, it is abundantly clear, were all in favour of the fowl.

The flintlock was invented in 1635. By the middle of the century breech-loading flintlocks, manufactured in Paris and of the most beautiful workmanship, were reaching Britain in some numbers. By the end of the century their use by British sportsmen had become general. In 1686 we find Richard Blome, in his *The Gentleman's Recreation*, advising on the choice of guns:

> Although those aforesaid *Lime-Rods* and *Intoxicating Baits* are very good for taking of *Fowls*; yet whereas their *haunts* and places of feeding are uncertain, which renders the said means uneffectual to attack the taking them by such like *Stratagems*; therefore the *Birding* or *Fowling-Piece* is of great use to those that are expert therein. And to be well accommodated therewith every *Fowler* ought to have them of several sorts and sizes suitable to the *Game* he designs to kill; for small *Birds* are with a small *Bore*, about four Foot, or four Foot and a half long in the *Barrel* is sufficient; for *Wildfowl*, as *Ducks*, *Herne*, *Wild-*

> *Geese* and the like, the *Barrel* ought to be about six Foot long, with an indifferent *Bore* under the size of a *Musket*.

You will notice that there is no mention of pheasant: the pheasant was then a rare bird in Britain. The partridge was common enough, but it would have been included in the category of small birds along with the pigeon which seems to have been more highly appreciated for the table at that time. It is evident that by Richard Blome's day not only had guns improved greatly but that ideas about their use had also advanced. Here is Blome again:

> It is now the Mode to *Shoot Flying*, as being by experience found the best and surest way; for when your Game is on the Wing, it is more exposed to danger; for if but one *Shot* hits any part of its *Wings* so expanded, it will occasion its fall, although not to kill it, so that your *Spaniel* will soon be its Victor; and if well disciplined to the Sport, bring it to you.

Ideas, however, seem to have been considerably in advance of performance. George Markland, whose *Pteryplegia or The Art of Shooting Flying* was published in 1727, held a pretty low view of the general standard of marksmanship in England.

> I have often wondered, why the *French*, of all Mankind, should be so expert at the Gun. I had almost said infallible. It's as rare for a progress'd Marksman of that Nation to miss a Bird, as for one of Ours to kill. But, as I have been since inform'd, they owe this excellence to Education. They are train'd up to it so very Young, that they are no more surprised or alarm'd with a Pheasant than a Rattle-mouse. The best Field-Philosophers living; for they are always there Masters of their Temper.

George Markland, who was himself a marksman of French excellence, did his best to educate his countrymen. His book is filled with sound advice –

> There sprung a single Partridge – ha! she's gone!
> Oh! Sir, you'd Time enough, you shot too soon;
> Scarce twenty Yards in open Sight! – for Shame!
> Y'had shattered her to Pieces with right Aim!
> Full forty Yards permit the Bird to go,
> The spreading Gun will surer Mischief sow;

But, when too near the flying Object is,
You certainly will mangle it, or miss;
And if too far, you may so slightly wound,
To kill the Bird, and yet not bring to Ground.–

as applicable today as when it was written.

This period – say, from the accession of the Tudors to about the middle of the eighteenth century – was really the 'golden age' of British field sports and also, as I believe, of British wild life.

Man hunted, it is true. Towards the end of the seventeenth century there were even a few fox-hunts, conducted more or less in the manner of today with mounted followers. Lord Arundel had a family pack of foxhounds in 1690 and the Duke of Buckingham another in Yorkshire at about the same time. The Duke of Richmond was hunting foxes in Sussex in the opening years of the eighteenth century: Squire Draper was hunting foxes in Holderness in 1726. All these were family packs hunting family land: neighbours rode to hounds only by invitation of the owner. Hounds were large and slow and a hunt was expected to last all day; though it was not, of course, the same fox that was hunted all the time. 'Compose your kennel of the biggest and slowest dogs you can get, respecting only cunning hunting and depth of mouth': it was hound-work and the music of hounds that were considered important, not the number of foxes killed. Foxes were then not nearly so plentiful as they are today. Squire Draper, a famous name in fox-hunting history, was much more concerned with the preservation of the few upon his land than with the killing of them. He was anxious to show off his hounds; their skill in following a scent, the beauty of their music. If foxes became a real nuisance, through raiding poultry or taking lambs, and had to be killed, then there were quicker methods than hunting to hand. It was easier to trap, to poison, to snare; to dig out an earth and knock the cubs on the head. And this was left to the estate owner's servants. Hunting was the recreation of gentlemen.

Shooting, with the gun, was also the recreation of gentlemen. Guns were expensive: far beyond the means of ordinary country folk. (It is not generally realized that the crossbow was still in widespread use among ordinary folk in rural Britain in the latter half of the eighteenth century, though it had probably wholly disappeared, save as a curiosity, at the beginning of the nineteenth.) And shooting, at this time, was a highly individual sport. A man, accompanied by his dog, went out by himself – at the most with one friend – on his own land, taking his chance with what he could find.

Throughout this period man, the hunter, shared the pursuit of game with the beasts and the birds of prey. And he did not resent their competition.

Chapter 7

THE AGRICULTURAL REVOLUTION

THIS virtually ideal (from the wild-life point of view) state of affairs came to an end with the Agricultural Revolution, which began in 1760 with the passing of the first Enclosure Acts. These private Acts of Parliament were designed to get rid of the open-field system. They were generally drawn in the same form, commencing with the recital that 'the open and common fields lie dispersed in small pieces intermixed with each other and inconveniently situated, divers persons own parts of them and are entitled to rights of common on them, so that in their present state they are incapable of improvement', and that 'it is desired that they may be divided and enclosed, a specific share being set out and allowed to each owner'. The men who promoted these Acts were wealthy country gentlemen. Parliament was controlled by them. There was, therefore, a minimum of debate and opposition. Enclosure Commissioners were appointed and their decisions had the force of law. Under their award the balks were ploughed up, the open fields and many of the commons – enclosures of common land continued for the first thirty years of Queen Victoria's reign – were divided into blocks for the several owners, and hedges were planted. Between 1760 and 1844 some 4,000 Enclosure Acts were passed, affecting more than 10,000 parishes and more than three million acres of land. The open-field system was totally destroyed. The whole face of the country was changed.

Now, there can be no doubt that these Enclosures were necessary and wise. The old open-field system – with its peculiarity of straggling or scattered ownership; the waste of time entailed in getting about from one part of a farm to another; the uselessness of one owner attempting to clean his land when it could be sown with thistles from the seed blown from the neighbouring strips of less careful and thrifty owners; the quarrelling about headlands and rights of way; and so on – was absurdly uneconomical. With enclosure each farmer became independent of his neighbour, was free to introduce improvements in crops and in animal breeding without fear that his efforts would be nullified by the prodigality of his neighbour. It was the Enclosure Acts that brought European fame as a farmer to Coke of Holkham: that made possible the work of Robert Bakewell in animal breeding. It was the Enclosure Acts that made it possible to keep herds of cattle alive through the winter, that freed men from the cheerless monotony of salted meat, that made the 'roast beef of Old England' famous throughout the world.

The Enclosure Acts were enforced by the Government: there was outright coercion. The peasant was not consulted and he was wholly unable to state his case. But it must not be thought that he was actually robbed of land: that was not so. A man deprived of pasture for his cow upon the common was given monetary compensation, though it would be only a few guineas. A man deprived of his strip in the open field might be given monetary compensation or an equal area of land somewhere else. His land or his right was never taken from him without some form of compensation. What he did lose – and for this there was no compensation – was his independence. Before Enclosure the men working the open fields had been farmers in their own right, working land their fathers and grandfathers had worked. It might be no more than an acre or two and it might be worked no more than two or three days in the week – the other days being spent working for some richer neighbour (that was the usual custom) – but it was his land. He was independent and he felt independent, the equal of his neighbour. With Enclosure that equality was removed. The few guineas of compensation were drunk in the inn. The equal acreage of land given, somewhere else and often some distance away, in compensation for the lost strip was useless to him, for he had not the capital to develop it. Perforce he had to sell to one of the big men: and to sell cheap. It is impossible to believe that this situation had not been foreseen by the promoters of the Enclosure Acts. That the Enclosures were necessary, that they benefited the national economy, is beyond question. They brought great wealth to the few (the great political peers), wealth to the county squire and the gentleman farmer – Coke of Holkham raised his rental from £2,200 to £20,000 a year in forty years – security and comfort (and often much more) to the tenant farmer. But they made of what had been a more or less self-supporting peasant class, with cows, pigs, sheep, geese grazing the common land, a labouring class with no stake in the land.

The Agricultural Revolution coincided with the Industrial Revolution. The Industrial Revolution was also, essentially, a rural revolution, since it changed the face of the countryside as greatly, and much more permanently. Indeed, in south Lancashire and the West Riding of Yorkshire, on Tyneside and the Clyde, in northern Cheshire and Derbyshire, in Staffordshire and Warwickshire, in south Wales – the countryside was not so much changed as obliterated altogether. The villages became towns and the towns spread and fused. Factories sprang up everywhere. Acre upon acre of moorland disappeared beneath row upon row of jerry-built terraces, built back-to-back and without water or drainage, and green valleys became blackened deserts. Lady Shelley, after a journey through the west midlands, wrote of 'that disagreeable, cold and manufacturing county which for twenty miles smokes from a thousand steam engines, so that at night the whole country from Birmingham to Wolverhampton appears to be on fire'. And that description would equally have fitted many other parts of the midlands and the north.

The Agricultural and Industrial Revolutions had an enormous effect upon wild life. The effect of both upon the human population has been very thoroughly studied and documented. Little is really known of the effect upon the wild life; and this is particularly true of the areas affected by the Industrial Revolution. In those areas men were too busy making fortunes or fighting for life to bother about anything else. But, obviously, as the towns spread, as vegetation vanished beneath bricks or was destroyed by smoke, as the streams became polluted, wild life was extinguished.

The Enclosures, on the other hand, must have had a beneficial effect on many, indeed upon most, forms of wild life. Visitors to Britain usually comment on our tiny fields and our miles of hedgerows; the patchwork of the English countryside. We, of course, take this for granted. It is difficult for us now to realize that there is virtually no hedgerow in Britain more than two hundred years old. But that is the fact: the hedgerows were planted by order of the Parliamentary Commissioners to enclose the blocks into which the old open fields were divided. Hawthorn was the plant most generally used, because it grew quickly and made a thick firm fence. The planting of the hedgerows must have made an enormous difference to many species of birds which had hitherto had some difficulty in finding nesting-sites. Improved agriculture made a great difference, too: feeding became much easier. England must suddenly have become a land of song. This is not guesswork. The fields on the Continent are huge and open; hedgerows are few and far between, their place taken by wire. Spring in agricultural France is silent compared with spring in England. Before the Enclosures England must have been so. And, of course, the great increase in the numbers of small hedgerow birds – and there must have been a great increase – would have meant a corresponding increase in the numbers of their predators.

But it was not only the birds to whom the hedgerows proved beneficial. It is from the planting of the hedgerows that we may date the great increase and spread of the rabbit. Before the Enclosures, though widespread, and plentiful in some areas, it was nowhere so plentiful as to be a nuisance. The hedgerows gave it shelter and opportunity to form extensive burrows in already loosened soil, and improved cultivation provided abundant food. Soon the rabbit was to become a menace to the farmer.

And it is from the planting of the hedgerows that we may date the spread of the brown rat. The brown rat first appeared in Russia from the east at the beginning of the eighteenth century and its spread westwards across Europe is very well documented. Indeed, exact dates are given for its appearance in some countries: a fact which prompted Dr Louis Sambon to suggest that they must have stamped their own passports! But the appearance of so formidable an animal is likely to have been noted by people already familiar with and frightened of (and with good cause) the smaller and weaker black rat. I do not think that the exact dates need be dismissed with scorn: I think they are likely to be accurate. It is, indeed, surprising that no more exact date than 'about

1730' can be given for Britain. The brown rat, as might be expected, was first noted in ports and it seems to have been confined to the ports and to the larger inland towns for the first thirty years or so of its residence in Britain. Indeed, I have been unable to find any reference to it as a country rat in Britain before 1762: and that reference comes from the immediate neighbourhood of Holyhead, a port. William Morris, writing to his brother Lewis from Holyhead in August 1762, says: 'On Monday last, I had three men reaping all my corn, and not a little trouble did we have; Norwegian rats devouring it standing.' Once it began to invade the countryside – and it did so, of course, from ports all around our coasts as well as from the larger inland towns – it spread with astonishing rapidity, exterminating the black rat as it did so. If there is anything to be said in favour of the brown rat, it is that it has virtually exterminated the black rat. The latter is now confined to a few ports only, and in these it maintains itself only by reinforcement from visiting ships.

I think that there can be little doubt that the brown rat's invasion of the countryside was facilitated by the planting of the hedgerows. There can be no doubt at all that it was the great increase in the numbers of rabbits and rats, following upon the planting of the hedgerows, that led to the great increase in the numbers of foxes. And it can surely be no mere coincidence that the establishment of fox-hunting as a nation-wide sport followed immediately upon the Enclosures.

The gamekeeper was also born of the Enclosures. The men who had been wont to take their game with lime, net or snare, and sometimes with crossbow, on the open fields and commons, not so much for sport as to eke out the family rations (especially in winter), simply could not understand that the enclosures of ground that had once been theirs made the wild creatures upon that ground private property. They found the hedgerows an aid to netting and snaring, and they continued to take their game. The men who had acquired property through the Enclosures took a very different view. They began to employ men to protect their game. These men had to know what they were about; to know the country and the people. They were, therefore, drawn from the ranks of the local poachers and were paid a small wage (but one slightly higher than that of the labourer) to stop their fellows poaching. What happened, of course, was that they did something to stop the poaching of others, but continued to poach themselves: 'the worst of poachers, taking one brace for the master, and two for themselves'. Black George in *Tom Jones* was typical of his class. It was only as shooting increased in popularity among the landed gentry (which it did very rapidly), only when it became an organized sport, only when a rent-free cottage and various other perquisites were added to his wage, that the gamekeeper became separated from the village folk and really concerned with the protection of game; not only from poachers, but also from natural predators. It is from this time – say, from about 1780 – that we may date the decline of the birds of prey in Britain.

It was the gamekeeper's job to increase the head of game for his master's pleasure. The most obvious way of doing this was to kill as many as possible of those creatures that preyed upon game. The gamekeeper of the late eighteenth and early nineteenth century really cannot be blamed for doing the job for which he was paid and housed.

It is more difficult to excuse the activities of many of the gentry of the time. During the first half of the nineteenth century the standard of marksmanship attained an excellence of which George Markland would thoroughly have approved: and this with the muzzle-loader. Now, the muzzle-loader was not an easy weapon. its action was slow, which meant that it was necessary to aim much farther in front of a bird than is the case with modern shot-guns. Moreover, reloading was a matter of time and, if done carelessly, of some danger. Yet men became extraordinarily proficient in the use of the muzzle-loader. For example, Coke of Holkham, who was not considered by his contemporaries as anything very exceptional as a shot, once killed eighty partridges in ninety-six shots. The real experts – like Colonel Peter Hawker, the author of *Instructions to Young Sportsmen in all that Relates to Guns and Shooting*, the acknowledged classic of the sport reckoned to maintain a rather better average day after day. Hawker once had twelve snipe in twelve shots; and the snipe is not an easy bird even for the best of modern shots armed with the best of modern guns. But Hawker was superb: it was a rare thing for him to miss any shot. It is easy to understand how men like these, walking their game up with dogs, would be likely to fire at anything that took wing; irrespective of what it was. Even Hawker, who was a good naturalist and knew his birds, has three bitterns and an avocet in his game-book. That, in a matter of fifty-one years of virtually continual shooting, shows a degree of moderation quite foreign to the times. Less knowledgeable men shot anything that flew, just because it flew. It was shooting, by gentry not gamekeepers, that brought about the extinction of the great bustard. Henry Stevenson in his *Birds of Norfolk*, writing of the sporting landowners of the last district in Britain in which the great bustard continued to breed, has this to say: 'Not a thought of the extermination of the species seems to have passed through their minds. Either they were entirely indifferent about the matter, or else they believed that since, as long as they could remember, there had always been Bustards on their brecks, therefore, Bustards there would always be.' I think this is true: what you see, you take for granted. I do not think that they were indifferent: I think the possibility of extermination never entered their heads. There was really no reason why it should. It is so easy to be wise after the event.

But there were others who did know what they were doing, and yet did it. Charles St John, who as a shot was not far removed from the Hawker class and who was a very good field naturalist, spent the summer of 1848 in Sutherland and, as he makes plain in his *A Tour in Sutherlandshire*, paid close attention to the osprey. Now, the osprey was disliked by gamekeepers because it is a bird of prey and hated by water-bailiffs because it takes fish in lake and stream. Both destroyed it because it was part of their duty to their employer to do so. St John had no such excuse. He killed for the love of killing, because he was a good shot and liked shooting, and because he was also a collector. He records that near Scourie he shot the hen of a breeding pair and took two eggs; that at Rhiconich he tried and failed to shoot the old birds, but took a young one from the nest; that at another nest near Rhiconich he shot the male, had a shot at the female and missed, and took the three young from the nest. Later he was told that the male at

Scourie had mated again, so he went back and robbed the nest. He ends this sorry tale thus:

> There are but very few in Britain at any time, their principal headquarters seeming to be in America; and though living in tolerable peace in the Highlands, they do not appear to increase nor to breed in any localities excepting where they find a situation similar to what I have already described (an islet in a lonely loch). As they in no way interfere with the sportsman or others, it is a great pity that they should ever be destroyed.

This, coming from Charles St John, who seems virtually to have wiped out the osprey population of Sutherland in a single summer, is rich indeed. And it must be remembered that there were many others of like mind – notably John Colquhoun, author of *The Moor and the Loch* – operating at the same time. It is all too easy, from this distance of time, to blame them, to say that they should have known better: all too easy to dismiss these sportsmen-collectors as maniacal murderers. In fact, they were men of their time: and their time was not ours.

Nor is it just to place (as is so often done) all the blame for the decline in the numbers of our birds of prey, and for the extinction of some as breeding species, on the gamekeeper and the sportsman-collector. That these two were responsible for much is beyond question: but responsibility was not theirs alone. Consider the case of the white-tailed eagle, now extinct as a breeding species and occurring only as a rare vagrant. The white-tailed eagle once had a range far exceeding that of the golden eagle – it nested in the Isle of Wight at the end of the eighteenth century – and at the beginning of the nineteenth century was much more plentiful than the golden eagle in the north. It was considered by shepherds to be a menace to sheep, a much greater menace than the golden eagle, and this view–presumably on sound evidence – received official support. The Commissioners of Supply, part of whose business was the encouragement of sheep-breeding, offered a reward of three shillings (a considerable sum in those days) for every eagle, white-tailed or golden, killed. The shepherds did not bother particularly about the golden eagle, but concentrated on the white-tailed. Some were shot, but, as the bird came readily to carrion, many were trapped and many more poisoned. Morris records that one hundred and seventy-one adult eagles were destroyed in Sutherland alone in a matter of three years. The reward was withdrawn in 1835. By that time the white-tailed eagle population had been exterminated on the mainland of Scotland (and there were probably no more than a dozen pairs surviving on the more remote cliffs of the outer islands), while the golden eagle had been confined to the north-western Highlands and so greatly reduced in numbers as no longer to constitute a menace to the sheep; if, in fact, it had ever done so. This, of course, was exactly the sort of situation to appeal to the ardent sportsman-collector. All collectors, whether of birds or stamps or whatever, wish to acquire the rarities. And this is particularly true of the professional collector, the man in the business for money. Alfred Newton, in his *A Dictionary of Birds*, has this to say of the extermination of the great auk:

> In Iceland there is the testimony of a score of witnesses that the latest survivors of the species were caught and killed by expeditions expressly organized with a view of supplying the demands of caterers to the various museums of Europe. In like manner the fact is incontestable that its breeding stations in the western part of the Atlantic were for three centuries visited and devastated with the combined objects of furnishing food and bait to the fishermen, and its final extinction was due, according to Sir Richard Bonnycastle (*Newfoundland in 1842*) to 'the ruthless trade in its eggs and skins'.

But if this was the age of ruthless slaughter, it was also the age of introductions. The red-legged partridge was introduced to Suffolk about 1770 and the capercaillie reintroduced to Perthshire in 1837. Both were introduced by shooting men to increase the variety of game in the district: were introduced to be shot. Both, particularly the capercaillie, flourished. It was also at about this time that the Japanese sika deer was introduced as a park ornament; to be followed, before the end of the century, by the muntjac and the Chinese water deer. All three have escaped and established themselves in the wild. The grey squirrel was also introduced at this time. The first authentic record is for 1876, when four were released in a park at Henbury in Cheshire, but there seems to be little doubt that the grey squirrel was present in Wales at a much earlier date. A letter written to the editor of the *Cambrian Quarterly Magazine* in 1830 states that

> ...in some retired glades of Montgomeryshire and Denbighshire a grey squirrel is found. Between Llanfair Caer Einon and Llan Eurvyl, in the former county, in a line nearly East to West, is a deep woody dingle, called Cwm Llwynog (the Fox's Dingle); its windings are intimately known only to the sportsmen, wood-gatherers, and peasantry of the neighbourhood, and the great retirement of the place accounts for the neglect its natural history has experienced; here a grey squirrel lives and breeds. The specimens I have seen were as large as a polecat, or a three-quarters grown rabbit; the head roundish, the eyes very prominent, the ears shorter than the common red squirrel's and not the slightest appearance of tufts upon them; the body and legs of a fine grey colour, the latter short and muscular and furnished with strong claws; there is a beautiful variegation of red along the sides of the ribs, from the elbow to the hind leg; the tail is covered with hair rather longer than in the common sort, and of a mixture of grey and black.... Regarding the habits of the grey squirrel, I confess my ignorance, further than that their nests present the same ingenious formation as those of the common species, and upon a natural presumption we may conclude that the grey sort also store up wild fruits against the winter...they are extremely shy on the approach of man, darting through the intricacies of the foliage with amazing swiftness, and taking single bounds of many yards length. The variableness of our climate considerably influences their season of eneration; but the young are usually produced from the latter part of April to the end of May. I have been informed that the grey squirrel monopolises the woods, and the common red kind are seldom seen near them, which appears reasonable enough, for the size and strength of the grey animal renders him more than a match for the other.

I have also seen a very fine stuffed specimen of the Welsh

grey squirrel in the possession of a gentleman residing in Chester; it was shot near Llandisilio Hall, Denbighshire, in October 1828.

Since it is evident that the author of this letter was thoroughly familiar with the red squirrel, there would appear to be no possibility of mistaken identification. This is not a description of the red squirrel in its greyish winter coat. It is a very good description of the grey squirrel. But precisely when the grey squirrel was introduced to Wales and by whom is not known. This colony seems to have died out, for there are no further records of the grey squirrel in Wales until 1903, when five were released in a park near Wrexham.

The grey squirrel has, since 1876, flourished so exceedingly that it has become a pest. It is customary nowadays to speak of the 'irresponsibility' of the landowners who first introduced it. It is all too easy to be wise after the event. These men had no more idea that they were releasing an animal destined to become a pest than had the man who first introduced the rabbit to Australia. He, indeed, was convinced that he was rendering his country a service.

Man has, of course, been responsible for a number of other introductions: a few deliberate, most accidental. Bennett's wallaby is well established on the moors along the Derbyshire–Staffordshire border. There should, surely, be no difficulty in controlling this animal should it show a tendency to spread. It may not, should they become well established, prove so easy to control the chipmunks which were liberated in Cheshire some three years ago. The accidental introductions have all been escapes from fur farms. It took a Government-sponsored campaign to rid us of the muskrat. The coypu and the mink are still with us, and are now so well established that they must be regarded as permanent members of our fauna. The mink is surely destined to become a pest: and, indeed, is already a pest in some districts.

But most of the recent additions to our fauna are birds: escapes from zoos or private collections. The Canada goose, the Carolina duck, the mandarin duck and the ruddy duck are all established in the wild as breeding species, though the distribution of the three ducks is still limited. All sorts of birds escape from collections of waterfowl from time to time, but they usually do so singly and so have no chance of establishing themselves. But in recent years small parties of ruddy shelduck and feruginous duck have been wandering about parts of the north and, though I do not know that they have yet bred, it seems not improbable that both will succeed in establishing themselves in due course. And birds also escape from aviaries. The budgerigar has bred in the wild on more than one occasion and the red-eared waxbill, a frequent escapee, may well do so in course of time.

To these accidental releases must be added one deliberate introduction: the bobwhite quail from America. This has been released in various parts of East Anglia in recent years for sporting purposes. It has established itself well and is showing some tendency to spread. Farming practice permitting, it might do as well as the red-legged partridge. Which brings me back to shooting and game-preservation.

About 1850 – no one knows the exact date and, oddly enough no one knows the inventor – the copper percussion cap replaced flint, steel, and pan. (It is interesting to note, in this connection, how many expressions in our everyday speech derive from the old muzzle-loader: 'a flash in the pan', 'keep your powder dry', 'as stiff as a ramrod', 'hang fire', 'misfire', and so on.) The copper cap was a revolutionary improvement. It was watertight – and this was a great step forward, for with the flintlock did it rain or come on to drizzle, and even in a mizzle, it was virtually impossible to keep the priming powder dry – and it was also the first step to the cartridge, which rendered possible the breech-loading principle. The original pinfire cartridge did not catch on – Colonel Hawker, though he claimed to have had a great deal to do with its development, was careful to stick to the muzzle-loader – because it was dangerous and caused some nasty accidents, most of them involving the loss of an eye. But the central-fire cartridge, which came into use in 1861, was an instant success. Landowners everywhere handed their muzzle-loaders over to their gamekeepers and took to the breech-loader. It was a revolution that was to have a profound effect on British wild life.

The breech-loader took the *energy* out of shooting. One has only to read Colonel Hawker's shooting diaries to realize that, in the fifty years they cover, he must have tramped many thousands of miles: and indeed, the Earl of Malmesbury, who also kept a shooting diary, estimated that in forty seasons he had covered more than thirty-six thousand miles. Shooting days were long and arduous. On 1st September 1845 Colonel Hawker was walking up partridges from 10 a.m. until 8 p.m. in a north-easterly gale. In those ten hours he bagged twenty-seven partridges and a hare; and he did not miss a shot all day. He was out again on the 3rd, again from 10 a.m. until 8 p.m., and shot twenty-nine partridges and two hares. (On both occasions Mrs Hawker accompanied him all day, and he makes it plain at the end of each entry that, while he was so dead-beat as to have no appetite, his wife was still 'fresh and hungry'. This is a very far cry from the conventional picture of the early Victorian lady in crinoline and poke-bonnet!) Grouse imposed an even greater hardship. The grouse-shooter reckoned to begin the day not later than 2 a.m., so that he might be on the moor at dawn, and he did not dream of giving up until dusk. Squire Osbaldeston, on 12th August 1828, 'a wild wet day', rose at 2 a.m., started to walk a rough and steep moorland at 3.30 a.m., and only gave up at 6 p.m. because there were then no birds to be found. On that day he shot twenty-two brace and was inordinately pleased with himself. And with every justification: to shoot forty-four grouse on a wild wet day with a flintlock was a very considerable achievement.

The breech-loader, with its ease and speed of loading, did away with all this. The guns could now stand in one place and have the birds driven over them: the walking, the really hard work, was done by beaters; men paid for the day's work. Walking up partridges still continued, but now it became an organized exercise rather than an individual sport. Four or five guns, with their attendants (somebody had to carry the cartridges) walked the farm crops in line, and close enough to ensure that all the birds in the field took wing and in such a manner that those not killed would take to cover in another field from which they could again be flushed, care being taken to see that they did not fly over the boundary into neighbouring property where they could not be followed. This is the method still in use today. But on all the large estates and on all the moors driving became the common practice. Driving of grouse and partridges commenced in the 'seventies: driving of pheasants (because the bird was still uncommon when the breech-loader came into general use) not until the 'eighties.

The evolution of driving demanded very much larger stocks of game. The gamekeeper necessarily had to become much more skilled, more scientific: a breeder of game as well as a preserver of game. It was in this period that the rotational burning of heather, to ensure new shoots for the grouse to feed on, began. It was in this period that the Euston system of partridge raising was developed. (In this system, named after the estates of the Duke of Grafton, the eggs of partridges are collected day by day by gamekeepers as they are laid, who substitute dummy eggs for them, returning the real eggs which have been brooded by hens or in an incubator when they are on the point of hatching.) It was in this period that the artificial rearing of pheasants was developed: the eggs being collected and hatched under farmyard hens on a rearing-field, the young birds, when nearly full-grown, being turned into the coverts which are to be their home and near which they are to be shot. (The driving of pheasants, by the way, is an art. Pheasants cannot be induced to fly strongly away from their home covert, but they will fly far and fast in order to get back to it. They have, therefore, to be quietly pushed, rather than driven, by a line of beaters away from their home covert and then, at a certain distance, stopped. The line of beaters then turns and flushes the birds so that they take wing. It may sound simple enough. It is, in fact, very difficult to do properly and pheasant-driving is now something of a lost art.) These new methods of rearing, coupled with driving in place of walking up and the breech-loader in place of the muzzle-loader, proved astonishingly successful.

Colonel Hawker, a superb shot, had only three days at grouse in his life and they were in November (not the best month) 1812. In those three days he shot sixteen and was well satisfied with himself. Squire Osbaldeston on the opening day, but a wild wet one, of the season of 1845 shot forty-four and boasted about it for the rest of his life. Both men were, of course, walking up and neither had a loader. On 30th August 1888 Lord Walsingham shot

1,070 grouse! (This still stands as the record for one gun in one day.) He was a very remarkable shot – he fired only 1,540 cartridges all day – but I doubt if he was really a better shot than Hawker or Osbaldeston. The difference lay in the method. His birds were driven over him by a line of well-trained beaters and he had a loader.

It was the same with partridges. Hawker, who specialized in partridges, shot 7,035 in the fifty-one years of his shooting career. His best season was 1828 when he shot 388, and his best day ever was fifty-six birds. On 8th September 1876 the Maharajah Duleep Singh shot 780 partridges. (This still stands as the record for one gun in one day.) Again, Hawker was walking up: the Maharajah had his birds driven over him and had a loader.

But the new methods made the greatest difference to pheasant-shooting. As I have said, in Colonel Hawker's day the pheasant was an uncommon, if not a rare, bird. In the whole of his shooting career Hawker only shot 575. In four years – 1825, 1837, 1843 and 1851 – he never so much as saw a pheasant on his Hampshire estate at Longparish, and in ten others the bag was but a single bird in each season. In 1895 at Highclere in Hampshire (no great distance from Longparish) no fewer than 10,807 pheasants were shot in three days by six guns.

The years 1880 to 1914 have been called 'the pheasant era'. During this period, on estates all over the country, pheasants were reared and shot in vast numbers. There are at least forty records of upwards of 3,000 pheasants being shot in one day. The actual record would seem to be 3,945 for 1913, but there are several totals which closely approach that figure. More important: there are a number of records of bags of upwards of 3,000 being obtained on successive days on one estate. For example, at Warter Priory in Yorkshire, 3,542 and 3,034 were accounted for on successive days in 1904 and in the following year, on the same estate and on successive days, 3,432 and 3,552. Bags of over 1,000 in a day were so frequent as to be commonplace.

In the same period, having regard to the different natures of the birds, there was a comparable increase in the size of bags of grouse and partridges. Between 1888, when Lord Walsingham staged his massacre, and 1915 there are five instances of bags in excess of 2,000 grouse being obtained in one day. The record stands at 2,929 to eight guns. In the same period bags of upwards of 1,000 in a day were not uncommon. Though bags of partridges in excess of 1,000 in a day were not common, the head of birds killed during a season on some of the better-known manors is significant. For example, 9,117 were shot on The Grange in Hampshire in 1887 and 8,512 at Holkham in 1905. It is also significant to note that one of the best shots in the country at the turn of the century reckoned to use 40,000 cartridges to his own gun annually!

Now, I am not interested in records as such. In any case, this sort of shooting bears no relation to sport as understood by such men as Colonel Hawker and Squire Osbaldeston: by comparison, it was mere target practice. I give the figures only because they are important as giving some idea of the enormous head of game that was raised by the new methods.

But raising game was only one side of the picture. It also had to be preserved. The gamekeepers did this by waging incessant war upon predators of all sorts and, indeed, upon any creature which might, just conceivably, prey upon game. The gamekeeper of this period has been well described by Eric Parler, drawing upon personal experience, in his *Shooting Days*: 'Every creature which was not game was for that reason the enemy of game, and so was to be shot or trapped or otherwise put out of a world which should only fitly contain partridges, pheasants, hares, rabbits, and gamekeepers. His traps befouled the woods; the hazels which he hung with stoats were horrors known afar'. This was the heyday of the gamekeeper's gibbet.

But again, let me stress that it is wholly unfair to blame the gamekeeper for his gibbet and the all too frequent slaughter of innocents. He was employed to raise a large head of game. He was employed to produce a thousand pheasants or more for the amusement of his employer and his guests at least on one day at the beginning of the season. And if he did not, he would be likely to lose his job. He could do no other than destroy what he thought might help to lose him his job by interfering with the amusement of his employer. The blame for what happened must rest squarely on the shoulders of the employer who, educated, should have known better.

What happened was this. In the middle of the nineteenth century the kite was nesting in every county but one in England, in every county but two in Wales, and was widespread in Scotland: by 1905 the total kite population of Britain had been reduced to five birds in Wales. In Colonel Hawker's day the hen harrier (a resident) was widespread on the uplands, and the marsh harrier and Montagu's harrier (both summer visitors) were to be found in most suitable parts of Britain: by 1907 the hen harrier was confined to the very north of Scotland and the other two had become very rare indeed. The honey buzzard, another summer visitor, was more common in the south of England than either the kite or the common buzzard in the middle of the last century: by 1885 only a pair or two were nesting in the New Forest. The honey buzzard is completely harmless to game. But it is a hawk, a large hawk, with a large hooked beak. And that was enough for the gamekeeper. At the same time the common buzzard and the raven, both greatly reduced in numbers, were driven into the more inaccessible parts of the west. Before the end of the pheasant era the polecat, the pine marten and the wild cat had been driven from England and survived only in small numbers in isolated localities in really difficult country. The fox survived: not because the gamekeeper had any liking for the fox, but because he was instructed to leave it alone by his employer who, if he did not hunt himself, was anxious to keep on good terms with the Master for social reasons.

One result of this wholesale slaughter of predators was a vast increase in the number of rats, rabbits, woodpigeons and grey squirrels: all creatures harmful to the interests of the farming community and one, the rat, also (as the gamekeeper was soon to discover) a more dangerous enemy on the pheasant-rearing field than fox, stoat and weasel put together. Another was a great increase in the incidence of disease, due to over-stocking and the survival of weakly birds which under natural conditions would soon have fallen prey to some predator. (The degree of overstocking was, indeed, often quite fantastic. In 1914, on the Witchingham estate of 4,125 acres, 4,021 partridges were shot. This gives an average of almost a bird an acre. The absolute maximum that good land can carry, if a healthy stock is to be maintained, is fifteen pairs per hundred acres. On very ordinary land (land comprising, say, 95 per cent of grass) the absolute maximum would be five pairs per hundred acres.) One is not altogether surprised to hear that the Witchingham birds were not strong fliers! Despite all the evidence stark before their eyes the great estate owners never realized that disease carried off more birds in a month than a host of predators would account for in a year.

Chapter 8

THE CLIMATE CHANGES

THE PHEASANT era came to an end with the outbreak of the First World War. Never since 1915 have pheasants been reared throughout the country on a scale familiar to the sportsmen of that era. The days have gone, and gone for ever, when on big estates four-figure bags were the order of the day for 'the first time through'; when at the end of the day a man would be hard put to it to recall a dozen of the hundreds of shots he had fired. The days of the 'record bag' – whether of grouse, partridge or pheasant – belong to history. And a very good thing, too.

Of course, pheasant-rearing began again at the end of that war: but, with very few exceptions (and none of those lasted very long), on a comparatively small scale. This was due, in the main, to the break-up of many of the large estates. The days when great landowners could afford to maintain huge stocks of game for their pleasure and for the amusement of their guests – often in so doing, and especially where pheasants were concerned, overriding the interests of their tenant farmers – had gone. With few exceptions, game was no longer preserved for pleasure but for profit. Guests tended to become 'paying'. With few exceptions, the owners of grouse moors – and of deer forests – now let the moor for the season, if possible to an American millionaire: if not, to a syndicate. The growth of the syndicate shoot was one of the features of rural life between the wars.

But to return to the pheasant. The war had shown the farmer that his work was easier and more profitable without a huge population of pheasants. There is an enormous literature on the food of the pheasant. If you read it carefully you cannot but come to the conclusion that the wild pheasant is omnivorous, taking vegetable matter and animal matter, with only a slight preference for vegetable matter. But that is the wild pheasant. The hand-reared pheasant, brought up on grain, is a very different bird. Lloyd George, opening his Land Campaign at Bedford in 1913, said:

> Here is one farmer who was sowing his crop – it was a field of mangels – the man assured me that there was not one mangel out of a dozen that was not pecked and destroyed by pheasants and that was a field that was clear of the plantations. Where you should have got thirty-five tons, here, you could not have had more than ten tons. It was not worth the labour and expense of carting. Labourers in the district said that the pheasants destroyed the produce of their gardens.

Well, we have all learnt by bitter experience that what politicians say rarely bears any close relationship to the truth. But here is Aymer Maxwell, a shooting man, writing in the same year of the damage done by hand-reared pheasants to farm crops:

> Except perhaps in the case of winter wheat, which remains a standing temptation, the bulk of the mischief is usually done in a short time. Fields of new sown or new sprouting wheat and barley, and grain between the ripening and the reaping, are only at the mercy of the wandering pheasant for at most few weeks of the year. Besides these cereals, among which at these critical periods of their growth a horde of pheasants can play havoc in a few days, patches of potatoes, standing crops of beans, and growing turnips are at times the subject of appreciable damage.

There can be no doubt that towards the close of the pheasant era the bird was doing considerable damage to agriculture. But this was really not the fault of the pheasant so much as that of the landowner or gamekeeper who had raised too large a stock for the normal food supply. The farmer knew this well enough. But on large estates, where pheasants were raised in their thousands, he was usually in no position to complain: he simply had to grin and bear it. Nevertheless, towards the end of the pheasant era, criticism was becoming more frequent, complaint more vocal. It seems unlikely that the pheasant era could have continued at full flush for much longer, even if the war had not intervened to swing the balance against the great landowners.

During the war there was no hand-rearing of pheasants. Food production was the important thing. The farmer learnt the lesson and was quite determined never again to allow pheasants in over-large numbers on his land. But he had also learnt that the wild pheasant, left to fend for itself, was on the whole beneficial to his interests because it destroyed an enormous number of wireworms and other noxious insects. The gamekeeper, by now a very different type from that described by Eric Parker, was at pains to drive home this fact. It is most revealing to study the correspondence columns of such papers as *The Field* and *The Gamekeeper* (both of which I have had the honour of editing) in the early nineteen-twenties and to compare them with those of the pre-war period. Prior to 1914 it was comparatively rare to find a gamekeeper putting pen to paper and when he did he was not concerned with the farmer. By contrast, *The Field* of 17th January 1920 contained a letter and photograph from H. Martin, head keeper to the Duke of Wellington, stating that he had examined the crop of a pheasant shot at Strathfieldsaye on 1st January and had found it to contain nothing but leather-jackets (the larvae of the daddy-longlegs), 702 in number, and that he hoped this would prove of interest to readers, 'especially the farmers, as we have the consolation of knowing that at least one pheasant has started the new year by doing some good for the crops'. In December of the same year *The Gamekeeper* published a photograph 'which should prove of interest to readers who are concerned in any form of agriculture'. The photograph showed 1,083 leather-jackets from the crop of a pheasant 'found trespassing upon some market gardens near Uckfield'. It will be seen that the pheasant, at this time, was thought to be in need of defence by the shooting fraternity.

Hand-rearing of pheasants did continue, of course, on the larger estates: but now always with a wary eye to the interests of the farmer. Elsewhere the old-fashioned rearing-field virtually vanished. Even those farmers who leased their shooting-rights were none too happy at the idea of having a rearing-field on their property, being fairly convinced that, no matter how well they leased the rights, they could put the ground to more profitable use. The game farm, specializing in the rearing of pheasants or in the sale of their eggs, really came into its own.

The syndicate shoot – new in the nineteen-twenties, but now commonplace – took many forms. At its best it operated then, as now, a rough shoot in wild or unfashionable country. In Wales, particularly, it was then possible to rent all the shooting-rights over several hundred acres, even perhaps a thousand, of rough land – some moorland, some mountain, a small tarn, some pasture, with maybe fifteen or twenty acres of arable – for as little as £50 per annum. This sort of rough shoot would provide plenty of game (none of it hand-reared) and some wildfowl in winter, and was very popular with small syndicates, half a dozen or so, of midland businessmen. No gamekeeper was employed, but it was always a condition of these leases – they are still obtainable, though at a rather higher rent – that once a month the members should hold a pest-drive. Syndicates of this sort did nothing but good, for they kept down pests like the carrion crow (which can take a fearful toll of mountain lambs), did a little to keep the rabbit in check, and at the same time provided the members with good, and very energetic, sport. (In the 'thirties it became the custom for these farmers to let the rabbit-rights to a professional trapper, separating them from the shooting-rights. This led, and rapidly, to a marked increase in the numbers of the rabbit. No trapper, having paid highly for the rights – which they were all eager to do – was going to trap himself out of a living. It was his aim not to clear a farm of rabbits, but rather to increase the stock and maintain it at a profitable level. For a man wise in the ways of rabbits, and skilful in the use of the gin-trap, this was easy enough to do.) On better land, though the shooting could still be properly described as rough, since walking-up was the normal practice, a gamekeeper was usually employed. Where the farmer merely leased the shooting-rights and took no further interest, was not himself a member of the syndicate, everyone gained. The game-keeper probably saw the members of his syndicate no more than once a week: he saw the farmer every day, had to keep on good terms with him and with his labourers. He probably hand-reared some game and he took care to see to it that his employers were satisfied, of course: but he also took care to see to it that the working of the farm was in no way upset. He struck a good balance – there was never any question of overstocking – and the countryside benefited. Some of the best-managed shoots I saw in the years immediately preceding the war were in the charge of single-handed gamekeepers on such farms.

But where a farmer – and this applied particularly to the 'gentleman farmer' in fashionable country – managed his own syndicate, had a gun in it, supplied the gamekeeper and maybe an assistant or two, the whole thing took on a very different complexion. Membership of syndicates such as these could be, and usually was, very expensive. (Membership of a grouse syndicate was even more expensive, which meant that most were made up of wealthy businessmen, climbing the social ladder: anxious to shoot grouse of course, but anxious to be able to invite the influential to shoot as guests and, perhaps as important, anxious to get a photograph of the marriageable daughter, tweed-clad and tramping the moors on 'The Twelfth', into one of the glossy magazines. For such privileges the business tycoon was eager to pay real money.) With the exception of the farmer, members of these fashionable syndicates had no interest in, no tie with, the land over which they shot. They gave nothing but cash, plenty of cash, and for their investment they demanded a large dividend in the way of game, expending as little energy as possible; which meant they required their game driven over them. This did not really make any noteworthy difference to farming and country life generally until about 1929. It is not generally realized now that the slump hit British agriculture pretty nearly as hard as it did British industry. But it did. It hit hard enough to make not a few of the 'gentleman farmers' realize that it was more profitable and infinitely less laborious to live off their syndicates than off their farms. To do so they had, of course, to provide their clients with a more plentiful supply of game. (In times of financial stringency people want more for their money.) Again there was massive over-stocking. But this time it was over-stocking on much smaller acreages; and that made it much worse. Inevitably there was disease. Between the wars there were at least four major outbreaks of grouse disease (strongylosis): that which occurred in the early nineteen-thirties being probably the worst in the history of the bird. In 1930 and 1931 there were also outbreaks of partridge disease (another form of strongylosis) which, in the midlands and the south of England, decimated the partridge population. At the same time there was an indiscriminate slaughter of birds of prey comparable to that at the peak of the pheasant era. The gamekeeper's gibbet, which in the early nineteen-twenties had been a relatively uncommon sight, became once again an all too familiar feature of the countryside. The lesson of the pheasant era, if it had ever been thoroughly digested by the shooting-man (which I take leave to doubt), had been wholly forgotten.

All this coincided with, or perhaps induced, a change in the climate of public opinion. I do not think there was any significant increase in the membership of societies advocating the abolition of field sports (then called 'blood sports'), though there was certainly a significant increase in the volume of propaganda, directed chiefly against fox-hunting, put out by these societies: sufficient, indeed, to lead to the foundation of the British Field Sports Society to combat it. But I do know that there was in the nineteen-thirties a sudden, spontaneous, and positively enormous increase in popular interest in birds. Bird books became fashionable reading, bird-watching a fashionable hobby: bird-baths, bird-tables, nesting-boxes *de rigueur* as part of the furniture of the garden. This vast army of bird-lovers, though for the most part completely unmoved by the often ill-informed and always too strident propaganda against fox-hunting and coursing, was deeply moved by the gamekeeper's gibbet, by the very thought of some birds being destroyed by man for whatever purpose: and made it abundantly clear that it was deeply moved. To begin with at any rate this was an unorganized, wholly spontaneous movement: and it carried all the greater weight because it was. The original reaction was solely against the gamekeeper and his gibbet. Later there developed a sustained attack on the egg-collector, the inspiration for which must, surely, have sprung from the Royal Society for the Protection of Birds. The R.S.P.B. has, from its foundation in 1889, always been an active and efficient organization, thoroughly knowledgeable and most level-headed, keeping its eyes on the stars but its feet firmly on the ground. It was the rare bird that was of the first importance at this stage and the prime enemy of the really rare bird was, undoubtedly, the egg-collector. If the inspiration did not come from the R.S.P.B., then it ought to have done so.

I was Editor-in-Chief of *The Field* at this time and, therefore, a natural target for anyone with anti-sport leanings. I was, indeed, thoroughly involved with both sides, for some of the pro-sport faction were of the opinion that I was not sufficiently whole-hearted in their support. I was immensely intrigued to note how eagerly shooting-men joined in the attack on the egg-collector. All the virulent propaganda against fox-hunting, to combat which it had been formed, had by this time left the British Field Sports Society serenely (and justifiably) confident that it could be virtually ignored, since most of it was so crude that it succeeded only in defeating its own ends, but it had been evident for some time that the attack on the gamekeeper was causing considerable perturbation in St James's Square. (The address, by the way, has long since changed.) As soon as the attack on the egg-collector got under way the British Field Sports Society produced a booklet, *Predatory Birds of Britain*, setting out which species it considered harmful and which it did not, and giving details of the measure of protection each species enjoyed at the time of publication. (The Bird Protection Acts in force before the war were entirely different from that in force today.) It is an excellent little booklet, but it is the foreword that is particularly interesting. I quote from it:

> To preserve game it is necessary to keep in check so-called 'vermin', both furred and feathered, and since feathered 'vermin' prey upon game, poultry and song-birds alike, the gamekeeper, in carrying out his duties to his employer, has been the most active agent in preserving in this country that chorus of bird song for which the British countryside is noted. It is perhaps significant that the National Trust, whose statutory duty includes the preservation of animal life, has stated in defending its policy of letting shooting on certain Trust properties that it does so *inter alia* to save the expense of 'keepering' and that 'the preservation of such life calls for active steps to keep down the number of species which may prey too heavily on other species'.

> But while, nowadays, credit is given gamekeepers for the part they play in preserving non-predatory birds, in the past aspersions have been cast upon them as a class as being responsible for the virtual extermination of some of our rarer birds of prey and for the unnecessary killing of others. To this broad assertion we should like to give our emphatic denial. Gamekeepers as a whole are students of nature, and kill only to protect their charges.
>
> On occasion, however, it is probably true that certain birds are shot in mistake for other species and without the knowledge that in some cases they are protected by law, but we believe that another and less known influence, which has no connection with gamekeepers or game preservers, has been the principal factor tending to thin out our rarer birds and to exterminate our rarest. We refer to the taking of their eggs by egg-collectors, for the prevention of which no stronger ally can be found than the gamekeeper.

That was jumping on the band-wagon with a vengeance! It is also, it must be admitted, a magnificent piece of special pleading: a wonderful example of how to skate over thin ice without actually getting your feet wet.

Now, I have no bias in favour of the egg-collector. I admit to having a strong bias, formed by close association over many years, in favour of the gamekeeper.

I do not pretend for one moment that the attack on the egg-collector was unjustified. On the contrary, I think that it was wholly justified and that it was high time it was made. But I must make it clear what I mean by an egg-collector. I do not include schoolboys, whose activities are mainly confined to hedgerow birds (most of which will lay a second clutch if the first is taken) and usually cease with their school days, if not before. Many a good naturalist has started his career by taking an egg from a blackbird's nest. It would be idle to pretend that the schoolboy does any real harm. It is the amateur kleptomaniacs and the professionals that matter.

I have no doubt that, before the change in the climate of public opinion, when the law was much more lax than it is today (and when nobody bothered over much about its enforcement anyway), the amateur kleptomaniac constituted a real, if relatively small, source of danger to the rare bird. I doubt if he does so today. Of course, simply by existing, he must constitute some danger. But he is an amateur. He has not the skill nor the cunning, nor in all probability the highly specialized equipment necessary, and is, therefore, likely to be caught. Today the penalties are so severe that I should be surprised if many members of this lunatic fringe survive.

The professional, however, is a very different kettle of fish. He is never himself a collector. He robs nests only in order to sell the eggs to wealthy collectors. (The owners of big collections rarely do any collecting themselves. Most have never so much as robbed a blackbird's nest and most would be quite incapable of climbing a tree. They collect eggs in the same way as other men collect stamps or coins or porcelain figures: by buying them.) The professional knows exactly what he is doing. He knows the risks and has weighed them against the reward. He is skilful and cunning, bold and ruthless, and – though his numbers have been greatly reduced through the efforts of the R.S.P.B. in recent years – he still constitutes, as he has always done, a real menace to the survival of the rare bird. So long as there are wealthy collectors, so long will there be men to supply their needs.

I have no doubt whatsoever that the professional nest-robber has brought about the extinction of some birds as breeding species in this country and has greatly reduced the chances of survival of a number of others. But it is as well to remember that no professional nest-robber has ever caused a plentiful species to become rare. The eggs are of no interest to him until the species is rare. For example, while the honey buzzard was plentiful in the south of England it was of no interest to the professionals. They became interested only when it became so uncommon that a couple of well-marked eggs would fetch five pounds. That was the standard price in 1860 for British-taken eggs of the honey buzzard. And, of course, the professionals took good care to see to it that the honey buzzard did not recover its old status. But they were not responsible for the original reduction in its numbers. Nor were they responsible for the reduction in numbers of the kite, the white-tailed eagle, the golden eagle, and others. That was the work of the shooting-man and the gamekeeper.

The change in the climate of public opinion had far-reaching effects. Although shooting-men joined eagerly in the attack upon the egg-collector, they also began, being well aware of the total implication of the change, to take tentative steps to put their own house in order. There was, immediately before the war, a noticeable decrease in the wholesale destruction of birds of prey, a noticeable increase in discrimination as between predator and predator.

This sort of discrimination is not always as simple as it may seem to the townsman, far removed from the reality of earning a living in the countryside. Predators do not only prey upon game. They prey also upon one another, and they also compete with one another. There are, moreover, many different forms of sport in this country and some of them have interests that are diametrically opposed. Furthermore, some have interests opposed, though never wholly so, to those of the farmer. For example, the sparrow-hawk is a most valuable ally to the fruit-farmer: it can be a menace on the rearing-field and to the poultry-farmer with young chicks. Again, the golden eagle is unquestionably a menace to grouse: unquestionably a valuable ally on the deer forest, where grouse themselves are a menace. The keeper on the moor and the stalker on the forest inevitably take a different view of the bird. Moreover, the hill farmer cannot be expected to view the loss of a lamb with equanimity. There are pros and cons attached to all predators; excepting always the rat.

Yet it was beginning to be realized that predators and prey had coexisted happily for thousands of years before man began protecting game. It was beginning to be realized that indiscriminate destruction of predators was a short-sighted policy, leading to disease and financial loss. More and more shooting-men began to accept the view that to maintain a *healthy* stock of game it was necessary also to maintain a stock of predators; that reduction rather than destruction was the correct policy. Some landowners began to afford protection to predators, at least to the less common, on their ground. For example, a number of owners and tenants of properties in Sutherland in 1939 agreed between themselves, without prompting from any society interested in preservation, not to trap or shoot the pine marten. No longer, over much of the country, was every peregrine falcon shot at sight. The golden eagle and the wild cat began to increase in numbers and even to extend their range. There was a movement towards some sort of balance.

Had not the war intervened I verily believe that a workable balance between all the various interests might have been achieved within a few years to the great benefit of the countryside.

Chapter 9

CHEMICAL WARFARE

THE SECOND WORLD WAR marked the end of an epoch. Man left the Age of Metals, in which he had been living for upwards of four thousand years, and entered the Age of Chemicals. He has been living in this new Age for less than a quarter of a century. He knows virtually nothing about it and has as yet, as is all too evident, not the slightest idea where he is going. He has been blinded by science.

Nowhere is this more apparent than in the countryside. And yet, at the end of the war, it did seem as though we were on the way, and that the way might be short, to a utopian countryside.

There was at the end of the war a great national fervour for social reform. Planning and Reconstruction were the watchwords. The bombed cities had to be replanned and rebuilt: their citizens given access to a countryside, large areas of which were to be preserved in all their beauty; safe from disfigurement by industry. So far as the countryside was concerned there was already a basis from which to work. Soon after its foundation the Royal Society for the Protection of Birds began to create sanctuaries on land which it owned or leased and to appoint wardens to guard the nesting-sites of rare birds on land which it did not by agreement with the owners or tenants. This, though very valuable, was necessarily rather piecemeal and was, in any case, concerned only with a comparatively small, though most important, element in the British fauna. In 1912, however, the Hon. Charles Rothschild founded and richly endowed the Society for the Promotion of Nature Reserves, which was concerned with nature as a whole. It was Rothschild who, in 1915, first produced a comprehensive list of proposed nature reserves in Britain. But very little constructive work was done until the late nineteen-thirties. Until then, the climate of public opinion, though not actively hostile, was wholly indifferent to the idea of nature reserves. By the end of the war the gentle breeze in favour of the protection of animal life had, fanned by a flood of books on all branches of natural history and a wealth of excellent and informed broadcasting, become a gale. Much the same thing had happened with regard to the preservation of the countryside. The Council for the Preservation of Rural England was founded in 1926, to be followed in 1927 and 1928 by the formation of similar bodies for Scotland and Wales. All three had already done much excellent work before the war. The stage was set and the time was ripe. There followed three famous Government reports – the Dower report of 1945 on *National Parks in England and Wales* (Cmd. 6628) which charted the course, the *Report of the National Parks Committee* (Cmd. 7121) and *Conservation of Nature in England and Wales* (Cmd. 7122), both published in July 1947 – which led to the foundation of the National Parks and National Nature Reserves under the National Parks and Access to the Countryside Act, 1949. The Act set up two bodies, the National Parks Commission (in England and Wales only) and the Nature Conservancy, whose responsibility it is to see that a number of areas shown on development plans receive special consideration within the planning system. (In June 1965 the Nature Conservancy became a component of the Natural Environment Research Council. The Conservancy undertakes broadly the same functions as it did before, but the formal responsibility now rests with the Council.) These areas include National Parks, Areas of Outstanding Beauty, Areas of Great Landscape, Scientific or Historic Value, Long-distance Routes, and national, local, and forest Nature Reserves.

The functions of the Nature Conservancy, as set out in its Charter, are as follows: 'To provide scientific advice on the conservation and control of the natural flora and fauna of Great Britain; to establish, maintain, and manage nature reserves in Great Britain, including the maintenance of physical features of scientific interest; and to organize and develop the research services related thereto.' These functions have been, and are, most admirably fulfilled. The area included within National Nature Reserves administered by the Conservancy is now well in excess of a quarter of a million acres, mostly in Scotland. These National Nature Reserves, which form the top rung of a long ladder and are the most important, vary enormously in size: some cover many thousands of acres, some only an acre or two. In addition there are Forest Nature Reserves which are managed for research and conservation by agreement between the Conservancy and the Forestry Commission and certain other major landowners. Then there are local Nature Reserves, which are established by local authorities on the advice of the Conservancy, wildfowl Refuges, and some two thousand Sites of Special Scientific Interest. These are sites which do not justify acquisition as National Nature Reserves, but which are of importance for the conservation of a habitat, a plant or an animal. They are scheduled by the Conservancy and notified to the local planning authority, which must then warn the Conservancy of any proposed development affecting the site. These sites are not, however, protected against any form of agricultural development nor against afforestation, and, in fact, even though the Conservancy must be notified, the only real obligation on the local planning authority is that it must take into consideration the interest of the site before determining a planning application.

In addition to all the above there are a number of important Reserves on National Trust property – notably Wicken Fen in Cambridgeshire and the Farne Islands off the coast of Northumberland – and a number of private Reserves, of which those owned or leased by the Society for the Promotion of Nature Reserves are particularly important. Finally, there are the Reserves owned or leased by the County Naturalists' Trusts, which are registered as companies limited by guarantee. The first of these, the Norfolk Naturalists' Trust, was founded as long ago as 1926. It was to be twenty years before Yorkshire followed suit. But in recent years the movement has gathered quite astonishing impetus, and has spread all over the country, so that now almost every county has its Trust and one or more reserves.

In 1958 the Council for Nature was established to act as the spokesman for all these national and local bodies, and also for a large number of natural history societies, representing in all more than 100,000 naturalists. It has done invaluable work in promoting and publicizing the cause of conservation – conservation as opposed to preservation: the maintenance and encouragement of habitats by wise use rather than their preservation through isolation as rural museums – and its monthly news-letter, *Habitat*, is unquestionably the most important regular publication concerned with British natural history.

All this forms a truly impressive network, from local level to national level, devoted to the cause of conservation. It is not surprising to find that Britain is widely accepted as having the most comprehensive and the most advanced system of nature conservation in the world. And that, indeed, is the fact. In no other country in the world is there so comprehensive a network. In no other country in the world has the cause of conservation so widespread, and indeed so passionate, a measure of public support.

And yet, in recent years, some ominous writing has appeared on the wall. The climate of public opinion certainly has not altered. On the contrary, the number of individuals who feel strongly about the necessity for conservation, the number of individuals who, though they may not fully understand the principles motivating the modern approach to nature conservation, yet have a deep interest in natural history, has undoubtedly increased enormously. But there can be no doubt that there has been some change in the climate of official opinion: that force of circumstance has somewhat dampened the revolutionary fervour that gave rise to the 1949 Act. Officialdom at all levels, from local authority upwards, is now subjected to pressures undreamed of by the reformers of the immediate post-war Attlee era: the pressure of population, pressures consequent upon that for all sorts of further development, financial pressures. In face of these the minority interest – and it must be admitted that, in terms of total population, natural history and nature conservation is very much a minority interest – has all too frequently either to give way or, if possible, to compromise. Far too much time has now to be spent by the Nature Conservancy and by the Council for Nature in trying to fend off, or in fighting, threats to important natural history sites all over Britain from one or another form of development. Many threats have been fended off: compromise has been achieved on many occasions. Some notable battles have been fought and won: some notable battles lost. The worst defeat – and a terrible augury for the future – has been that suffered over the reservoir now being built at Cow Green in Upper Teesdale. This reservoir, built to satisfy powerful commercial and political interests, will drown a botanical site of international importance: a site around which a National Nature Reserve had been created in order that it might be protected for all time. A truly shocking precedent has been set.

That, ominous though it is from the point of view of the future of some National Nature Reserves, is not, from the point of view of conservation as a whole, the most ominous writing on the wall. More dangerous, because less obvious, are the enormous changes in modern farming practice.

Farms are tending to become larger and larger. There are today arable farmers in East Anglia who maintain that a holding of one thousand acres is the smallest viable unit. Already there are those who maintain that these men are out of date and that nothing less than two thousand acres is really viable. If we ever become members of the Common Market, it is certain that farms everywhere, and not merely arable farms, must become larger.

It is not only the farms that are becoming larger: it is also their fields. Modern machine farming demands large fields: the larger the better, because the easier and the more economical to work. And so the hedgerows are disappearing. Many miles of hedgerows have already disappeared in East Anglia (particularly in Lincolnshire) and elsewhere. An expert, at the Oxford Farming Conference in 1965, pointed out that the destruction of forty-eight miles of hedgerow in Lincolnshire would reduce costs by as much as £1,440 a year. That sort of figure must make a strong appeal to a farmer. Many more miles of hedgerow will inevitably be grubbed up: and, indeed, the policy has official blessing. Whether it is a wise policy is another matter. Soil erosion, hitherto a matter of comparatively little importance in Britain, has increased a great deal in East Anglia in recent years. In the Thetford neighbourhood, in the great gale of 1955, vast quantities of surface soil were blown hither and thither across the countryside. The high winds of March 1968 had an even greater effect. Mr P. R. P. Ingram of the School of Biological Sciences, University of East Anglia, writing in *The Times*, said that 'the Yare valley below my window is filled with a brown cloud of blown topsoil, presumably eroded from the fields of central Norfolk'. That letter produced another from Mr H. N. Walters pointing out that soil erosion 'is very evident in Leicestershire, Nottinghamshire, and Lincolnshire. In the local paper last week it was reported that the Grantham to Lincoln road was partially blocked and made dangerous for traffic by sand drifts'. It cannot be questioned that the removal of hedges and small 'unproductive' coppices from the countryside has soil erosion as a natural physical consequence. There can be no doubt that in the long term the destruction of hedgerows is an unwise policy. But who, now, thinks in the long term? Anyway, the arable farmer now really has no choice. Modern machine farming demands large fields.

There is another aspect of large-field arable farming which, though less obvious to the casual eye, is every whit as important and no less ominous from the point of view of nature conservation. Large fields, covering many acres, often several hundred acres, and maybe devoted to a single crop, are impossible to weed and are too extensive to allow insect predators and parasites to exercise satisfactory control over insect pests. Large fields demand chemicals, sprays and fertilizers, in large quantities.

Now, there can be no question whatsoever of the great value of chemicals in agriculture. I quote from Garth Christian's *Tomorrow's Countryside*:

> The achievements of British agriculture in the past quarter of a century are remarkable by any standards. Only ten years ago farmers were congratulating themselves that the national average yield of wheat had substantially increased to 24 cwt to the acre; they now accept without surprise a national average harvest of 30 cwt for wheat and 29 cwt for barley; and there are some farmers, in the words of Dr G. W. Cooke, of Rothamsted Agricultural Experimental Station, who think that 'if wheat comes out at less than 45 cwt and barley at less than two tons, something is wrong'.

That is precisely what has happened. Of course, all sorts of things have helped – better farm machinery, higher-yielding and disease-resisting strains, and so on – but it has, beyond all doubt, been the revolutionary advances in the application of chemicals that has been the chief contributory factor to this astonishing achievement. And it is reasonably certain that yields will increase yet further within the next few years. No one would deny the value of chemicals in agriculture. It is their effect beyond agriculture that is causing so much heart-searching.

Leaving aside the chemical fertilizers (which, it must be remembered, are as beneficial to weeds as to crops), the chemicals that matter are the herbicides and the insecticides.

The average man, when he thinks about (that is, if he thinks about it at all) the effect of toxic chemicals on animal life, thinks only in terms of insecticides. In fact, herbicides have a much greater, much more widespread, effect on animal life than do the insecticides proper. (The fact has been tacitly acknowledged by the manufacture of the term 'pesticide', which is now universally used, to cover herbicides and insecticides.) The herbicides may not be designed specifically to kill insects – and, indeed, are not – but they do do so; and in the most wholesale manner. For the herbicide, in controlling or killing its weeds, is removing the food plant of some insect or perhaps of many different kinds of insects and thus causes their death, if not directly by poison, then indirectly by starvation. Even when the weed is not actually the food plant, its disappearance, though that disappearance may be only temporary and local, may have, indeed must have, some effect on some aspect of insect life. For example, charlock, dandelion and rosebay willow-herb are notable sources of nectar, coltsfoot, poppy and sowthistle notable sources of pollen, for bees. The weeds of one man may well be the treasure of another. Some weeds are of immense significance for the bee-farmer since they may fill a gap between more important sources of supply. And, quite apart from honey, we all know how important a part bees play in the rural economy. But, considering weeds simply as food plants that are essential for the maintenance of some part of the insect population, it is obvious that their removal must interfere with a food chain, the extent and value of which we do not know and cannot, therefore, fully appreciate.

Of course, there are selective herbicides: the so-called selective weed-killers. I have said 'so-called' deliberately. 'Weed-killer' is a misnomer. The word 'kill' implies finality; complete victory, no more weeds. No selective weed-killer gives final victory. And no weed-killer ever will; for, if it did, it would kill all other plants as well. The selective weed-killers are, in fact, selective deterrents: controllers rather than annihilators.

And very efficient they are. The arable farmers of East Anglia, using selective hormone weed-killers, virtually rid their fields of charlock and poppy. They were naturally delighted. But it was not long before they found that their place was rapidly being filled by more persistent weeds; and particularly by the wild oat. The wild oat is a most invasive weed and was soon infesting hundreds of acres. And, since it is closely related to the crop, its control posed a problem. The problem has been solved, of course: but at a cost to the farmer of about £3 an acre a year. Higher yields more than justify that expense.

But there are risks attached to this sort of thing. It has already been shown that some plants are able to breed resistance to certain chemicals used for their control. The best-known case is that of *Erechites hiercacifolia*, a common and persistent weed of sugar plantations, which has bred resistance to 2,4-D; so that, now, a much stronger chemical has to be used. It seems probable that it is only a matter of time before other weeds also breed resistance to the chemicals used for their control, which will mean the development of still more powerful agents. This has, in fact, happened with insecticides.

Before the war the insecticides most commonly used – it is difficult, now, to realize that, then, insecticides were not commonly used on a large scale – were derris and pyrethrum. Both are of vegetable origin: derris powder being obtained by grinding the root of a South American plant and pyrethrum being prepared from the flowers of *Chrysanthemum cinerariaefolium*. Neither is really dangerous to mammals or birds – if dangerous at all – and pyrethrum is a particularly effective insect-killer. Unfortunately it is rather expensive to produce and so is little used in agriculture.

The insecticides which have been most widely used since the war fall into two groups: the organo-phosphorus compounds and the organo-chlorine compounds. The first group contains some very poisonous chemicals, like parathion and schradan, which necessitate the wearing of protective clothing by the operator, and the complete closing-off of the area treated for as long as six weeks for the most lethal members of the group. Such precautions were not always taken, of course. Parathion, in particular, caused the deaths of many farm animals, large numbers of birds, and some humans. So far as I am aware, parathion is now used only on glasshouse crops and on peas. When the toxicity of such compounds was realized – and parathion is by no means the most toxic – research work was directed to finding compounds less toxic to warm-blooded animals, but still lethal to insects. The chemists came up with malathion, which was for a time thought to be completely harmless to all but insect life. And then there were a few human fatalities. These were found to be due not to malathion itself (by itself it is harmless enough) but to the fact that the man using it had recently been using parathion or some other organo-phosphorus compound. There was, in other words, a cumulative effect. There can be no doubt that without proper care – and all the manufacturers of these compounds are careful to point out the protective measures that should be taken – organo-phosphorus compounds, either singly or in amalgamation, are dangerous. But that they are most effective insecticides cannot be questioned. And they also have this great advantage: they lack stability, are not persistent, soon break down into quite harmless substances.

This is not the case with the much more widely used organochlorines: BHC, DDD, DDT, aldrin, dieldrin, endrin, heptachlor and the rest. The best known of these is, of course, DDT. The use of DDT during the war undoubtedly saved many thousands of lives by controlling insect vectors of disease; and, particularly, lice and mosquitoes, carriers of typhus and malaria. There were then those who thought, and at that time with every justification, that the answer to insect-borne diseases had been found. It was then thought, and it continued to be widely believed for some ten years after the war, that the perfect insecticide – almost instantly lethal to insects, absolutely harmless to man – had been discovered. We now know that all this was over-optimistic.

A number of insects, for example, have developed resistance, bred immunity, to DDT. And these include those notorious carriers of disease, house flies and lice and mosquitoes. A number of agricultural insect pests have also developed resistance. Insects breed in enormous numbers and have, therefore, all the more chance of developing immunity. So far as the application of insecticides is concerned, this is constantly happening. You have but to study the long lists of 'more efficient' chemicals produced year after year to realize that something must have gone pretty awry with the previous year's batch. The fact is that, in this field at any rate, the scientist is never one jump ahead: he is always trying to catch up. The chemical used this year to obtain satisfactory control of some insect may well next year, or at some time in the not-distant future, prove to be inadequate. Nature has a way of hitting back: and hitting back hard.

An outstanding example of this is provided by the brown rat, man's inevitable companion and most formidable enemy. For years past the most widely used rat poison in the world has been 'warfarin', an anticoagulant. Until quite recently no one has ever had the slightest doubt about its efficacy. Now, in Scotland and Wales and in the west midlands, in Denmark, and at the same time, far away on the other side of the world, in Guyana, the brown rat has developed immunity to warfarin. Welsh rats have been fed a diet of warfarin and nothing else for months on end, and have suffered no ill effects whatsoever. In Britain (and it is the same elsewhere) these immune rats are spreading from the centres in which the immunity developed, handing on their immunity. It is not a situation which can be regarded with complacency. It is an object lesson in the resilience of Nature.

But to return to the organo-chlorine insecticides. So far as man himself is concerned, it is not their actual toxicity that is important. It is true that endrin is as poisonous as the more highly poisonous organo-phosphorus insecticides – endrin, however, is but little used in this country – but aldrin, dieldrin and heptachlor are not much more, if any more, poisonous than DDT. A man could kill himself by swallowing DDT: no doubt about that. But he would have to swallow an awful lot of the stuff. So far as he is concerned it is not the toxicity of these insecticides that matters: it is their persistence. The Agricultural Research Council, referring (in 1964) to the use of these organo-chlorine compounds as insecticides, had this to say:

Small accumulations and residues found in human bodies, human food, wild birds, fish, soil and some of the creatures which live on it, water, suggest a widespread contamination of 'living things' and their environment.

I daresay that, if you think about that, you will find it pretty horrifying. Now consider this quotation from Garth Christian's *Tomorrow's Countryside* (1966):

> ... the concentration of DDT in human fatty tissues has risen significantly during the past fifteen years, and in 1963 the average figure for Americans was 12 ppm.* In a random analysis of restaurant meals all contained DDT. However, when Dr W. J. Hayes fed human volunteers with relatively large amounts for over a year, no ill effects were noticeable. During 1963, when the average concentration of DDT among English and German citizens was 2 ppm, it was discovered that dieldrin, too, was now stored in the fat of residents in southern England. More recently, in 1965, tests carried out jointly by the Ministry of Health and the Laboratory of the Government Chemist showed that human fat from post-mortems in widely scattered parts of England and Wales contained an average of 3·3 ppm of DDT and 0·26 ppm of dieldrin. Nineteen samples of breast milk taken in London, Leicester and Cardiff all contained minute portions of DDT, BHC and dieldrin.

He goes on:

> There are some distinguished scientists who fear that if man continues to absorb these substances, which are now known to affect the fertility of raptorial birds, he may one day face a human tragedy far more dramatic and disastrous than the thalidomide affair. Still more scientists dismiss these fears. It would be fair to say that at the present time no link between pesticides and ill health has been proved. Nor is there any evidence that the residues have any effect on human fertility.

* parts per million.

All that is perfectly true. There is, as yet, no evidence that these residues damage human health. But there is incontrovertible evidence of a significant increase in the level of residue in humans. And I know of no evidence to suggest that this is unlikely to continue. How high a level is acceptable? What will constitute a dangerous level? If you ask the scientists these questions what you will get is an evasive answer. If you ask enough of them, you will, I think, come to the same conclusion as I have: that they do not know. And, indeed, they do not. As an American scientist, Robert L. Rudd, has pointed out in his *Pesticides and the Living Landscape*, 'hazard from ingestion of residues over very long periods is yet to be assessed. Should residues increase generally, any hazard would be magnified. Moreover, there is now no escape from these residues for most of the population.'

And that is perfectly true. There may be no danger – nobody knows – but there is certainly no escape. The whole world has already been contaminated. In the Antarctic, Adelie penguins, Weddell seals, fish and skuas have been found to contain insecticide residues: in the Arctic, insecticide residues have been found on lichens and in caribou, and are doubtless present in other animals. Insecticide residues have also been recovered from rainwater and from the air. I am not suggesting that these residues have been found in such amounts as to be biologically significant. That is not so. But it is surely significant that these pesticides should have reached the remote ends of the earth: that they are present, no matter in how small quantities, in places where man has never used them. Here is evidence enough of their persistent and invasive quality. Indeed, there is now no escape.

When these organo-chlorine insecticides were first introduced it was not realized that they possessed this quality of persistence. In a remarkable and most moving speech at the first 'Countryside in 1970' Conference, Dr E. F. Edson of Fison's Pest Control, one of the most distinguished chemists in the world, frankly admitted that 'it came as a great surprise to those of us who have an interest and experience of this problem'. There were other surprises. I quote from *Pesticides and Pollution* by Dr. Kenneth Mellanby, Director of the Nature Conservancy's Monks Wood Experimental Station:

> A reason for the great persistence of some of these substances is that, instead of breaking down to give more or less inert substances, as do so many pesticides, these are transformed in the soil or in living tissues to equally or even more toxic chemicals. Thus heptachlor turns to heptachlorepoxide, a stable substance considerably more poisonous than heptachlor itself. Aldrin disappears quite quickly, and early workers thought that this meant an end to its potency; however it was soon found that it was transformed into dieldrin, which could be traced unchanged in soil for ten years or more. It was only later found that dieldrin could be stored by living animals, mainly in the fat deposits, and that it remained fully toxic and able to cause poisoning when once more it was mobilized.

Not a very pleasant thought!

It must not be thought that aldrin, dieldrin, heptachlor and the rest were placed on the market without first being subjected to detailed and extensive toxicity tests. Every conceivable care was taken and over a long period. The members of the Association of British Manufacturers of Agricultural Chemicals, all highly qualified men, many with international reputations, have always been most conscientious in this respect, being well aware that they might be wielding a two-edged sword. But – to quote Dr Mellanby again –

> ... toxicology is a difficult subject. Different species differ in their susceptibility to different poisons, and it is at present impossible to foretell with certainty the effects of a substance on one species by trial using another. Furthermore, the same species may be more tolerant of a poison at one season of the year than at another, and the way in which the poison is administered affects the results. No insecticide can be tested against every species of bird and mammal which may be at risk.

Note the words 'at present'. What I said at the beginning of this chapter is true. We have entered a new Age. We have been living in it for so little time that 'at present' we know virtually nothing about it. Those two quotations, surely, amount to a frank admission of that fact: an admission that, no matter how carefully conducted the experiment, no matter over how many years it is conducted (and some have covered several years and many generations of laboratory animals), no matter how stringent the precautions to prevent damage to other forms of wild life and to domestic stock and to humans, there is always the possibility of changes in the substance itself once it reaches the soil or after a period in the soil, and always the possibility of side-effects that cannot 'at present' be foreseen by even the most brilliant scientist. Dr Edson has put it bluntly: 'A chemist can unwittingly create in one year a biological problem so complex that his biologist colleague may need a decade to evaluate or solve it.'

Of ordinary people I think it was the angler who, finding his fish dead or dying in what had hitherto been a clean stream and still appeared to be, first began to suspect that chemical sprays did more than they were designed to do. But it was undoubtedly bird-lovers who, writing to the Press in the spring of 1956 (the year after aldrin and dieldrin first became available to farmers) about bird deaths, first began to arouse public alarm. In the spring of each of the following years increasingly large numbers of birds of many species were found dead, especially in East Anglia, and the number of letters of protest steadily increased. They were dismissed as the sentimental outpourings of nature lovers. The official attitude was that there was no cause for concern. Then came the great 'fox death' of the winter of 1959–60 – no fewer than 1,300 foxes were actually found dead in East Anglia, and there were others elswhere – and at the same time the Royal Society for the Protection of Birds and the British Trust for Ornithology began investigating bird deaths, publishing the first of their Reports in 1960. It soon became apparent that the chief cause of death, both of birds and mammals (and the mammals included foxes, badgers, dogs, cats and an otter), was chemically dressed seed and the chief agent was dieldrin. No one knows how many birds were killed by poison over the country as a whole up to the end of the spring of 1961 (which was probably the worst year; though in the spring of 1960 over 5,000 birds, most of them woodpigeons, were picked up on one farm of a thousand acres), but the number must have run into many thousands.

In that spring (1961) at Tumby in Lincolnshire 6,109 birds of more than fifteen species were picked up in a period of twelve weeks, at Anglesey in Cambridgeshire 4,000 in a period of ten weeks; and there were other counts as high or almost as high. The total death roll, even in the places where careful counts were made, must, of course, have been much higher than the actual number of bodies collected. Much would depend on the amount of time the observers had to spare – dead bodies disappear remarkably quickly in the countryside – and, in any case, it is all too easy for even the most highly trained observer to overlook a small body.

Yet in that same spring of 1961 we find Sir William Slater, F.R.S., former Secretary of the Agricultural Research Society, writing in the *New Scientist*: 'On the farmlands of these crowded islands food production must come first. We cannot sacrifice the efficiency of our food production in order to avoid the loss of a few birds and mammals killed by chemical, a loss which is only a minute fraction of those due to natural causes or the guns of sportsmen'.

No one can disagree with the first sentence of that quotation. The rest is balderdash. It may be true (I do not know) that the loss was only a minute fraction of that due to natural causes in the course of a year: but, even if you accept that, it was an additional loss and, moreover, one concentrated in the most important period of the year, the spring, the breeding season. As for the guns of sportsmen: sportsmen do not shoot chaffinches, a species which suffered badly. But let us be fair. Let us consider a species which has received attention from the guns of sportsmen. The sparrow-hawk successfully survived persecution – continual and highly professional persecution – at the hands of the gamekeepers of Britain from the beginning of the pheasant era onwards. In 1955, after three-quarters of a century of continual persecution, it was the most common and the most widely distributed hawk in Britain. In 1957, only two years after the introduction of the organo-chlorides, it had completely vanished from most areas of Britain. Today it is a rare bird, fully protected throughout the year.

Sir William Slater represented the official attitude at the time. It was the work of three voluntary bodies – the Royal Society for the Protection of Birds, the British Trust for Ornithology, and the Game Research Association – backed by insistent pressure from the Nature Conservancy, a number of members of all parties in Parliament, and a great weight of public opinion that brought about a change of heart at the Ministry of Agriculture. A voluntary agreement was reached between the Ministry, the insecticide manufacturers and the farmers that dieldrin should not be used to dress spring-sown wheat, but that it might be used in autumn. Dieldrin is by far the most efficient protection against the wheat-bulb fly and that pest is most common in autumn. It does also occur in spring, of course, though not usually to so great an extent, and control can then be successfully achieved with BHC (which has not yet been shown to be seriously harmful to birds, though it is a highly toxic non-selective insecticide), but not so successfully as with dieldrin; so the farmer made a very considerable sacrifice. Manufacturers and farmers loyally abided by this agreement, and the result was really quite remarkable. There has been no further mass slaughter of birds in spring and some species, such as the chaffinch, which had been severely affected, more or less re-established themselves. Following the recommendations of the Cook Committee in 1964 further voluntary restrictions on the use of aldrin, dieldrin and heptachlor were agreed by the manufacturers and the National Farmers' Union to come into force at the end of 1965. This was a magnificent example of public spirit on the part of the farmers. Dieldrin is by far the most efficient insecticide for controlling carrot fly (a pest which causes very considerable financial loss), while as a sheep-dip it is invaluable since one application gives protection from 'strike' for a whole summer. There are other methods of controlling carrot fly which, if not quite so effective, are successful enough: but for the sheep farmer the ban was irritating and very expensive, both in terms of money and of labour.

It was during this period that a great decline in the numbers of birds of prey first became really noticeable. The case of the sparrow-hawk has already been mentioned. Now it was noticed that kestrels, particularly in East Anglia, were dwindling in numbers at an alarming rate and that the peregrine falcon was fast disappearing everywhere. In neither case could the gun be blamed.

Despite the gun and the egg-collector, the peregrine population remained remarkably stable up to the outbreak of war in 1939. During the war a large number were shot – the chief prey of the peregrine falcon is the pigeon, and homing pigeons were used on a large scale during the war – and by 1945 there was but a small population left. Shooting, except for an occasional bird taken by a gamekeeper, stopped after the war and by 1955 the peregrine falcon population was back to pre-war level: a wonderful recovery. And then came the crash. By 1962 the population had been reduced by at least half. A large number of adult birds died of poisoning – as was to be expected since the favourite prey, the pigeon, was the species most susceptible to dieldrin poisoning – but it was also found that many mated birds either failed to lay at all or laid infertile eggs or took to egg-breaking. The same phenomenon was noticed in golden eagles, kestrels, buzzards and various members of the crow family. In 1964 two scientists of the Nature Conservancy reported that the number of pairs of golden eagles rearing young over a wide area of the western Highlands had declined from 72 per cent in 1937–60 to 29 per cent in 1961–3. Since then sixteen of the thirty-nine pairs are said not to have laid and of the twenty-three that have laid about half have broken their eggs. Unhatched eggs from seven eyries have been examined and were found to be contaminated with dieldrin and other insecticides. I think there can be little doubt – the evidence is surely overwhelming – that dieldrin and other insecticides can cause sterility or a grave disturbance of nesting behaviour in birds of prey. And this is not confined to Britain – the same thing has been noticed in the United States and on the Continent – nor is it confined to the land birds of prey and the carrion-eaters. The osprey, newly returned as a breeder to Scotland, lives virtually entirely on fish; yet one clutch of infertile Scottish eggs was found to be contaminated by insecticides. And this has also been found to be the case with the heron and with at least a dozen species of seabirds.

A bird in as perilous a situation as some of the birds of prey is the common partridge. Indeed, in so parlous a situation is it that the word 'common' no longer applies. Over considerable areas of England the partridge is already extinct or on the verge of extinction as a breeding species. This is not due to disease or to over-shooting. It is due entirely to modern farming practices: to the use of insecticides which contaminate the seeds eaten by the adults and which destroy the insects which are almost exclusively the diet of young partridges during their first three weeks, to the use of herbicides which destroy the food plants of many insects and a fair proportion of the food of the adult birds, and to the destruction of hedgerows and the burning of stubble. It is ironic to recall that for more than a century it was commonly said that 'good farming and partridges go hand in hand'. Since the partridge is principally a bird of agricultural country and light soils, since it tends to avoid areas of intensive cultivation and needs some rough ground for cover and feeding, its future over much of Britain would appear to be bleak indeed.

Public attention has naturally been focused on the insecticides because their effects are often so immediately and dramatically obvious. The effects of herbicides (which account for a large percentage of all the agricultural chemicals used in Britain today) are often less immediately obvious. Of course, no less than the ornithologist, the botanist knows what is happening. But it has proved very much more difficult to arouse public opinion, on a national scale, over the loss of a plant than over the loss of a bird. Nevertheless, public opinion has often been aroused by the action of local councils in spraying roadside verges. Justifying such action, councils always emphasize the need to keep down vegetation at the sides of roads in order to ensure better visibility for the motorist. One may be forgiven for believing that the prime reason is not care for the motorist but the saving in labour costs. Be that as it may: the fact is that much of the beauty of British roadside verges (and, too often, of canal towpaths also), the dense growth of coarse grasses and flowering plants – cow parsley and hogweed, convolvulus and creeping thistle, speedwell and buttercup, nettles and thistles, and the rest – is blackened every spring and probably at least once again in summer.

One result of this has been the disappearance, over large areas of the country, of many butterflies and moths. It must be all of ten years since I saw an orange-tip, perhaps as long since I saw a hedge brown; butterflies which, the one in spring and the other in summer, were once an essential part of the British wayside. They have vanished from a neighbourhood in which they were once common because their habitats, the food plants of their caterpillars, have been destroyed. Fifteen years ago I could leave the bedroom window open and the light on and be certain of attracting many moths: now I regard myself as singularly fortunate if I get half a dozen in the room in a night.

Most herbicides, apparently, break down in the soil fairly rapidly. But it cannot be too strongly stressed that we really have no knowledge at all (for the simple reason that we have not yet been using them long enough) of the long-term effects on the soil of these chemicals. It may well be that by saving ourselves trouble we are storing up trouble for ourselves. The scientists certainly do not know that we are not.

Chapter 10

THROUGH A GLASS DARKLY

I AM no less qualified than the greatest of scientists to look into the future. That statement will not commend me to any scientist. Nevertheless, it is a statement of fact. The scientists know what they are doing today. They know no more about tomorrow than you or I. They would not be human if they did.

Of one thing, however, we can be quite certain. Pesticides have come to stay. Man, it is true, has done perfectly well without them for thousands of years. But, having got them, he cannot now live without them. What he has yet to learn is to live with them. And what that really amounts to is that he has to learn to live with himself in a new world of his own making.

We live on what is naturally one of the most beautiful islands in the world. That is a fact which we are always forgetting. There is another fact, even more important, which we are always forgetting (which, perhaps, many people never realize) and that is that our scenery, the landscape as we see it today, has been largely, indeed almost entirely, fashioned by man. This is as true of the hill and moorland pastures of the north and the west as of the arable and grass fields of the east and and the midlands and the south. For the large area of uncultivated land in the north and west, even though it has never been ploughed and manured, has been burned and grazed for centuries: the vegetation that covers it is 'natural' only in the sense that it has not been deliberately planted. It is only on a few mountain-tops, only on a few stretches of remote coast, only in a few (very few) isolated valleys, that nature in Britain is really 'virgin'.

The story of the fair face of Britain can be summed up in half a dozen words: axe, crook, chisel, billhook, plough and scythe. It has been the minor craft of agriculture that have moulded the British landscape, have fashioned British scenery.

As we have seen, it took man a very long time – roughly four thousand years: from about 2400 B.C. to about A.D. 1840 – to remould the landscape of this island. In the course of that time he exterminated those creatures hostile and dangerous to himself and his family – the bear, the lynx, the wolf – and in so doing he was perfectly justified. No one would pretend otherwise. In the course of that time he exterminated, as breeding species, a number of birds by destroying their habitats; and particularly by draining and claiming marshland. No one would pretend that he was not justified in so doing. Throughout that time he strove unceasingly to control the weeds in his crops: throughout that time he trapped or shot the animals that ate his crops. No one would pretend that he was not justified in so doing. Throughout that time also he greatly increased the numbers of some birds by his agricultural activities and, latterly particularly, by the planting of hedgerows. Throughout that time man was well aware that he was a guest in the land and that his stay was short. During all those years, from the moment he became an agriculturist, he ceased to live solely in the present, just from day to day as does the hunting animal. He cared for the land, put back into it what he took out, tried to leave for his successors earth in as good, if possible in a better, condition than when he himself took it over. The great landowners and the squires, the farmers and the simple cottagers, possessed a deep sense of trusteeship for the land they owned or used. And they made very few mistakes.

'Man finds a forest and leaves a desert.' The charge is true: and true of man all over the world. But it has become true of man in Britain only quite recently. We know, for the evidence is stark before our eyes, the damage that man has done in Britain since about 1840. 'The evil that men do lives after them'. And it is, unfortunately, all too true that there are very few places in the world in which commercial greed has done more harm – or at least is more noticeable – than here in Britain. To quote Shakespeare again: 'The evil that thou causest to be done, that is thy means to live.' The lust for quick profits has left some shocking memorials behind it. Yet these shabby monuments to ruthless business enterprise still amounted to no more than spots upon a fair face up to the outbreak of the Second World War. Since that war they have become a rash and that rash must, inevitably, spread and thicken.

Man could not have made Britain, naturally beautiful with its immense diversity of geological strata, so much more beautiful were it not a small island. There is no town, no village, in the whole of its length and breadth which is more than three hours' motoring from the sea. But this small compass, once so great a blessing, has now become a millstone around our necks. Our population is increasing so rapidly that it has been estimated that a quarter of a million acres of good farmland will have been converted to urban use by 1970: that a further two million acres will have been swallowed up by towns by the end of the century. But that is not all. To serve this enormous urban development new roads and motorways, new airports and reservoirs, with their attendant services, will have to be constructed. By the end of the century, only three decades hence, the demand upon land is going to be well in excess of two million acres.

It is against this background of steadily increasing urban spread and steadily decreasing agricultural acreage in a small island that one must consider the future of British wild life.

The vast majority of the British population were born in towns and live and work in towns. The most tragic consequence of the Industrial Revolution has been that the town has become separated from the country: that the townsman and the countryman now live in worlds apart, having no true understanding of each other's way of life.

Not that the townsman has lost interest in the countryside. Far from it: as urban life becomes more and more regimented, is dictated more and more by the clock, as work becomes more and more mechanized, as individual skills and the pride that goes with them give way to the monotony of the conveyor belt, so is the townsman's longing for the peace of the countryside increased. The widespread interest in natural history is one symptom of this: the great interest in gardening another. British suburban gardens are the most beautiful, the most lovingly tended in the world. It is significant that the successful townsman secures for himself a week-end cottage in the country or removes himself to some semirural area and commutes at the first possible opportunity. And it is surely significant that each week-end millions pour out from the towns and cities to seek relaxation in the countryside. The pity – and it is here that the divorce between town and country is most noticeable – is that most do not recognize the countryside as the workshop of those that live in it, but regard it as the playground of those that do not. Nevertheless, all this is to the good and must be encouraged to the full.

The demand for rural recreational facilities will increase steadily in the next three decades and is going to impose its problems. Already the National Parks, covering an area of some five thousand square miles in England and Wales, are subject to pressure. And so are some of the National Nature Reserves. I do not know how many people now visit the skiing slopes on the Cairngorms, but the figure will be in the neighbourhood of thirty to forty thousand. Already one finds waste paper, cigarette cartons, bottles and tins in many a burn high on the mountains. The number of skiers will certainly have doubled, possibly trebled, by the end of the century. What will conditions be like then? Consider the Broads. More than three hundred thousand people now visit them annually. That figure may well have doubled by the end of the century.

The Cairngorms is a National Nature Reserve of some forty thousand acres, which was established in 1954 to guard the most important relict fauna and flora of inland Scotland. Access to a part of it is restricted in autumn. The Broads constitute the greatest and most valuable reservoir of wetland fauna and flora, including many rarities, in Britain. Within the area there are three National Nature Reserves and one Nature Reserve of the National Trust. (If special biological interest is taken into account there should surely be at least seven others.) Access to these four Reserves is by permit. This is an excellent system now. Will it work at the end of the century, by which time it is estimated that the population will have increased by seventeen million? I am very doubtful. It is difficult to see how restricted access, how access by permit, could be enforced in the face of national pressure, which will then, inevitably, be backed by powerful commercial interests. We have already seen what happens to a National Nature Reserve in face of commercial pressure. And it is as well to remember that the vote that matters is the urban vote.

I have no doubt that there will be a still wider interest in natural history at the end of the century. Some wonderful work is being done in this direction in schools, on radio and television, by means of 'nature trails', and so on: and some of it will stick. I think that there can be no doubt that awareness of the need for conservation of nature, active interest in the idea of conservation, will have developed greatly by the end of the century. But it will still be a minority interest. We delude ourselves if we think that it will not.

But, for the whole future of wild life in Britain, even more important than recreational and commercial pressures upon National Parks and National Nature Reserves is the problem posed by agriculture. Indeed, this is the vital problem. For there can be no doubt that with a greatly reduced acreage and a greatly increased population farming will have to become more and more intensive. There is no escaping that fact: no escaping the fact that this brings us back to pesticides.

Pesticides have come to stay: and rightly. We may now find all sorts of faults, in the way of side-effects, with DDT. But the fact remains that DDT has saved millions of lives and will save millions more. And there is no getting away from the fact that as agriculture becomes more and more intensive, so pesticides will become more and more essential as a protection – even if often no more than a temporary protection: the chemists have learned that insects are likely always to have the last laugh – against the ravages of pests. That fact must be faced. And, for the nature-lover, it is an unpleasant fact.

The Royal Society for the Protection of Birds has restored to the British breeding list four or more species which had been lost to us. These birds – the osprey is by far the best known because it receives the most publicity – are now safeguarded, sometimes at very considerable expense, in their breeding haunts. They are safe from the gun, safe from the egg-collector, those two dreaded villains of the nineteenth and early twentieth centuries. They are not, as has been shown in the case of the osprey and the golden eagle and others, safe from the chemist.

Now, it is all very well for the bird-lover, concerned about the reduction in the numbers of our birds of prey and deeply anxious to see the osprey firmly re-established as a breeding species – but safe in his suburban home, his town flat, not dependent upon the land for his living – to cry 'Thou shalt not'. If he had to earn his living as a farmer he would almost certainly take a very different view of pesticides. And it must be said that so far all the concessions have come from the farmer and the manufacturer. For example, by agreeing not to use dieldrin as a sheep-dip the sheep farmer added greatly to his expenses and enormously to the labour of his shepherd. (It is, of course, true that the sheep farmer got along well enough before the advent of dieldrin and it might be said, therefore, that he has not made a very big sacrifice in the interest of the golden eagle. This is a false argument. One might as well say that man got along well enough before the advent of the motor-car and that, therefore, he would not be making a very big sacrifice if he agreed to a ban on the use of motor-cars in the interest of clean air.) If the farmers and manufacturers make concessions now, then the conservationists must not be surprised, and certainly will be in no position to complain, if, later on, they have to make concessions to recreational and commercial interests: as surely they will.

But the present pressing problem is that of pesticides. The scientist is no longer entirely happy about his miracle. It would not be an over-statement to say that, having found that insects have the ability to breed immunity to certain poisons, the medical entomologist is very worried. Dr James Busvine, a leading authority on tropical medicine, has said in this connection that 'it is difficult to exaggerate the seriousness of this problem, which threatens to vitiate the great progress being made by modern insecticides in stamping out insect-borne disease'. I doubt if the agricultural chemist is much happier: for, in attempting to solve one problem, he has given birth to a host of others and to a considerable volume of public protest. (Most of this springs, as a study of the literature reveals unmistakably, from fear: not from fear for the future of wild life as such, but fear for the future of the human race. And this in an age in which man threatens his own existence by the production of nuclear bombs!)

Dr Edson, who is most sensitively aware of all the many problems associated with the use of pesticides, speaking at 'The Countryside in 1970' conference, said (as his final sentence): 'Above all other I feel that we need some continuous mutual understanding, not only of the difficulties of conserving wild life, but also of the difficulties of trying to retain control of that unwanted wild life which is a menace to our human population.' Born of that conference, and particularly of that speech, there is now a joint committee of ABMAC (the Association of British Manufacturers of Agricultural Chemicals) and those interested in the conservation of wild life. I do not know that there is continuous mutual understanding – that would, perhaps, be asking for too much – but there is continuous contact: and that, in itself, is a great step forward. The Chairman of this committee has said: 'Responsible people have become fully aware of the need to avoid indiscriminate destruction or harm to species which endow the countryside with beauty and interest. By a collaborative effort between those who manufacture, distribute and use pesticides and those who are primarily concerned with the conservation of wild life, new pesticides are being developed that are less persistent. Moreover, codes for the safer use of existing pesticides are being widely adopted in this country.' One might almost be listening to a Cabinet Minister soothing the voters!

Of course, responsible people are aware of the need to avoid indiscriminate destruction, etc. The point is *how* to avoid this while still using pesticides. It is nice to know that less persistent pesticides are being developed. But how long does it take before one can be sure that a pesticide is less persistent (less persistent than what?) and what happens in the meanwhile? And does less persistent mean less lethal? If it does, then the reaction of the ordinary farmer will be to double the dose. If it does not, then there must inevitably be destruction and harm to species other than those it is intended to control. Nature is indivisible: a fact which we overlook, but of which our ancestors were always aware.

Poison is not, of course, the only means of controlling pests, insect or other. The gun and the trap alone, or in combination, have never succeeded in controlling, let alone exterminating, a pest. But, when used in conjunction with these two, poison can sometimes prove most effective: as we have seen in the case of the white-tailed eagle, a bird of prey which we would now be delighted to welcome. On the other hand, the rabbit defied gun, trap, snare and poisoning by gas pumped into the burrows until 1953. In that year, however, myxomatosis reached Britain and the rabbit rapidly succumbed. Myxomatosis is a virus disease and is, therefore, a form of biological control.

The disease was deliberately introduced into Australia, where the rabbit was a very serious pest on grazing land (an adult rabbit eats about half as much as a sheep), on a number of occasions prior to 1950 and failed to take hold. In 1950 it did take hold and with the most dramatic results. The death rate was virtually 100 per cent, and the stock-carrying capacity of sheep farms was greatly increased. Although the rabbit in Australia has since bred some immunity to the disease, the numbers have been so greatly reduced that it is now possible to ensure that it will never again be allowed to attain pest status.

At about the same time as the Australian introduction of 1950 myxomatosis was deliberately introduced into France with equally dramatic results. In the autumn of 1953, apparently accidentally, it reached Kent and, despite attempts to contain it, rapidly spread all over the country. The mortality rate was well in excess of 90 per cent and in many areas the whole stock was wiped out. For the first time farmers realized what the rabbit really had cost them.

Prior to 1953 it was estimated that the rabbit cost the country at least £50,000,000 a year. One would have thought that myxomatosis, particularly as it is specific to the rabbit and affects no other animal, would have been unreservedly welcomed. It was, and indeed is, by the farmer – it is illegal to spread it deliberately, but there is undoubtedly a black market in infected rabbits, for some degree of immunity has now been bred in certain areas – but it is a most unpleasant disease to look at and there was a public outcry against it; mostly arising from those with no direct stake in the countryside. We are a strange people. I can recall no public outcry against the use of poisonous gas against rabbits. But, of course, the poisoned animal died out of sight. If myxomatosis had been specific to rats, rather than rabbits, I doubt if there would have been any outcry at all. But the rabbit is a pretty little animal with a little white tail: and that made all the difference.

Pesticides are not, of course, the only means of controlling insect pests. Biological control – a method unlikely to commend itself to the manufacturing chemist – has proved most successful in many parts of the world. Work at the Glasshouse Crops Research Institute at Littlehampton has shown that the red mite, one of the most damaging pests of glasshouse crops, can be so successfully eliminated in the course of one season by the introduction of a Chilean predator that no further treatment may be needed for a number of years. The cost of this control is put at about one-tenth of that of chemical control.

Dr F. J. Simmonds, Director of the Commonwealth Institute of Biological Control, in a recent lecture to the Royal Society of Arts, stated that in seven projects the Institute had secured permanent control of pests which had already operated for up to thirty-three years. The total outlay on these projects was £25,000. He gave some dramatic figures of results. For, example, control of a sugar pest on St Kitts had yielded a return in terms of increased sugar output of £1,600,000: and this for an outlay of £200! Dr Simmonds admitted that biological methods had not always been successful and that there was then a short-term advantage in chemical control: the pest does not build up immunity, no poison is used and so there is no possibility of the indiscriminate killing of beneficial insects (such as bees) and no possibility of harm to other forms of wild life (such as birds of prey): and since there is no poison, there is no possibility of harmful residues being left in the soil or in water, no possibility of harmful residues building up in food chains which may ultimately reach man.

It seems to me that today, in this country, we hear far too much about pesticides; far too little about the possibilities of biological control. It may be that biological control on any large scale, other than in glasshouses, is impossible in this country. I do not know. (I do know that I use no pesticides in my fairly large garden and that my garden does not suffer from this lack of chemical attention: but I realize that it might well be different on a hundred acres of one crop.) It may be that we have already gone too far:

that the British farmer, now educated in pesticides, would not consider anything so homespun as Nature's way. Yet I find it difficult to believe that the British farmer is really less intelligent than the sugar planter in St Kitts or the farmers of California or the Soviet Union, in both of which countries biological control of certain pests has been tried with great success.

In the meanwhile, pesticides are here and here to stay for the foreseeable future. In view of that, none but the most starry-eyed can be optimistic about the long-term future of some forms of British wild life. But, though one cannot but be somewhat pessimistic about pesticides – it is impossible not to be, for, on all the available evidence, the insect pests and the brown rat seem likely to be the only creatures to benefit in the long run – it would be most unjust to lay all the blame for the disasters that have been, and for the disasters which, one suspects, are yet to come, on the agricultural chemists and their poisons. They are, after all, but the men of their time: just as St John and Colquhoun were the men of theirs. It is the times, not the men themselves, that are awry. Almost as important as the poisons are the changes in farming practice, change in rural pattern, above all change in human attitudes: the loss of the deep sense of trusteeship for the land owned or worked.

Not all the changes in rural pattern are harmful, of course. The planting programme of the Forestry Commission, for example, is highly beneficial to woodland birds and to the fox; and, in certain districts, to the badger, the polecat, the pine marten and the wild cat. But there is one change in farming practice which, especially when coupled (as it is and as it will continue to be) with the use of pesticides, is bound to have a far-reaching and ever more damaging effect on wild life generally, but particularly on many species of birds. I refer to the destruction of hedgerows, now encouraged by Ministry grants, to allow for ever larger fields. (I wonder how long it will be before that same Ministry will be forced to make much larger grants to the very same farmers to encourage irrigation to compensate for the erosion encouraged by its hedge-removal grants?) The Royal Society for the Protection of Birds spends a great deal of time and money (and rightly) taking steps to ensure the protection of rare birds as breeding species. But what is to become of the common hedgerow-nesting species – blackbird, song thrush, hedge-sparrow, yellowhammer, whitethroat and the rest – as the distance between hedgerows widens, as the variety of insect food that is so essential to the rearing of young birds (and not only the young of the professional insect-eaters) becomes more and more difficult to find? Already their numbers are diminishing in agricultural country in East Anglia and elsewhere.

As at this moment of time, peering into the future, it would seem to be that only in gardens with trees and shrubs to provide nesting-sites, with lawns and borders to provide optimum feeding conditions, only in gardens where pesticides are used sparingly and with discretion or not at all, that such birds – or some of them, for maybe the yellowhammer and one or two others will never take kindly to the proximity of houses – will be able to find sanctuary. By the same token, it may be that the partridge will survive in agricultural country only as a golf-course bird. The fox has already discovered that suburban life has many advantages – freedom from hunting, freedom from shooting, the freedom of the dustbin – and is steadily building up a large population in stock-broker territory.

I am not so concerned about the future of our rare birds; for the simple reason that they will always be rare. No amount of protection is going to make the avocet or the black-tailed godwit – let alone the osprey – common as breeding species in Britain. The requisite numbers of suitable breeding-sites are not now available nor ever again will be. The birds of prey, as we have seen, face special hazards. But, once re-established as breeding species, such wetland birds as the avocet and the black-tailed godwit are now afforded such excellent protection in their breeding haunts by the Royal Society for the Protection of Birds that there would seem to be no reason to fear for their future in small numbers. The future of many of our common birds, it seems to me, is much less secure. And if you think that is over-pessimistic, then consider the present status of many of our butterflies which, a mere twenty years ago, were common and widespread.

Of course, with few exceptions, all wild birds are protected throughout the year by the Protection of Birds Acts of 1954 and

1967. One might think that that should be enough to ensure their future. Under these Acts it is an offence for any person to kill, injure or take, or attempt to kill, injure or take, any wild bird: to take, damage or destroy the nest of any wild bird while that nest is in use: to take or destroy an egg of any wild bird. You could not wish for anything more comprehensive than that. Our Bird Protection Acts are, indeed, the most comprehensive in the world.

Nevertheless, these Acts are born of the past – of the bad old days of St John and his like – and are aimed essentially at the shooting-man and the egg-collector. The shooting-man is no longer a menace: and the few professional egg-collectors are too well known (and the nesting-sites of rare birds too well guarded) to be a serious menace. These Acts, based on situations and events now long past, are not really relevant to the present. We shut our eyes to the facts if we believe that they are. It is true that section six of the amending Act of 1967 (which has now been incorporated in section five of the principal Act of 1954) specifically prohibits the use of poisonous substances. Well, what are pesticides?

I make the point not to suggest that any farmer should be prosecuted for using toxic chemicals on his land and thereby causing the deaths of birds and mammals – that, obviously, would be quite ridiculous – but to point the difficulty of framing legislation for the protection of one form of wild life under modern conditions.

Change in the past, as I have tried to show, has been gradual. Man's hunting activities, whether for food or sport, have never seriously affected wild life: only recently, when they have been conducted for commercial gain, have they done so. But from Neolithic times onwards his agricultural activities have always affected the pattern of wild life. But always, until very recently, there has been time, any amount of time, for plants and animals to adjust. The tempo of life was slow: so slow that man was aware that he was part of nature, that he was as much subject to her rhythms as the wild plants and animals. He was fully aware that nature is indivisible. Plants and animals and man lived in balance.

That is no longer so. The pace is no longer that of the horse, but that of the jet. From time immemorial his religious leaders have been telling him that he is a being set apart from the rest of creation, the supreme being. Now, he really believes it. Now, he really believes that he is the supreme autocrat with the right to control, and even to exterminate, competitive forms of life. Man has separated himself from nature.

Fortunately, he is illogical. Having developed enormous powers of destruction, having reached the point at which he may exterminate himself at any moment, he has developed a deep and genuine desire to conserve what is left of wild life, both plant and animal, around him. Much, far too much, has been lost. But there is yet much to be saved. The problem is how to conserve it in a small island with a rapidly increasing population; a small island in which already almost all the land and water is managed, and must continue to be managed, for economic reasons.

Time, in the jet age, is not on the side of the conservationists. It is not enough to plan for the immediate future; to plan for the 1970s. Time moves too quickly. The immediate future is the present and that present past before plans can be translated into action. That has happened with our rural policy time and again in the last twenty years. If we are to conserve our wild life – more important, if we are to conserve wild-life habitats – then we should be planning now, not for the next decade, but for that time, thirty years hence, when our population will have increased by seventeen millions. It may be said that as we cannot know the full extent of the problem then, we cannot find the solution now. But we can at least make an effort not to be overtaken by an event which can already be foreseen.

Mr John Coleman-Cooke, Honorary Secretary of the Council for Nature, speaking at the Council's 1968 Annual General Meeting, said: 'In a progressive society it is now taken for granted that straighter and wider roads are more important than "wilderness", that larger airfields must swallow up the countryside. It may be expedient for the national economy that such decisions should override conservation objections; but is it wise in the long-term?' It must be remembered that, without a flourishing national economy, there will be no long-term at all. Conservationists really must bring themselves to accept the inevitable now, and try so to influence planning that nature has some chance of living in harmony with man.

In that same speech Mr Coleman-Cooke said: 'To sell the concept of wild-life reservoirs as biological investments is a tough assignment in a market that is invariably concerned with relatively short-term values.' I confess that I am not sure precisely what is meant by 'biological investment'. But I am sure that the whole concept of 'wild-life reservoirs' is already out of date.

Of course, there must be some nature reserves. Coastal reserves, such as those at Minsmere and Havergate, designed to protect the breeding haunts of some rare form of wild life, are essential and are so situated that there would seem to be no reason why they should not continue to operate for ever to the great benefit of man and nature.

But, looked at coldly and without sentiment, the inland reserve – in the sense that it can be shut off and isolated, reserved for some particular purpose – is doomed. In so small an island as Britain it is already, even with the population at its present figure, impossible to isolate inland areas. Some authority may say that this or that inland area is of national importance because it contains rare plants or for some other reason, and should therefore be preserved. And that may well be true. But no matter where that area is, no matter what steps authority may take by restricting access or whatever to protect it, it is at the mercy of outside and hostile influences, agricultural or commercial. As we have seen, even with the most comprehensive Bird Protection Acts in the world, it is impossible to protect birds from the ordinary everyday activities of agriculture and commerce.

Dr Mellanby has said: 'Our need is for a more positive approach on the part of the whole nation, not just a few enthusiasts often thought of as cranks, to the problem of wild-life conservation in our crowded island'. And that is the bald truth. The future of the wild life of Britain does not lie in a few reserves, but in man's attitude of mind.

The vast majority of wild creatures live in the workaday countryside. Nature conservation, if it is to be successful, must form part of a national policy which includes agriculture and industry, urban development and transport. It cannot be separated from these things.

If we do not wish to end up with a vast population of pesticide-proof insects, such parasites on man as the brown rat and a few garden birds, then we must re-learn, and quickly, the fundamental truth that nature is indivisible.

BIBLIOGRAPHY

A NUMBER of the publications listed below may appear from their titles to bear no relation to wild life in Britain. All are, however, essential reading for anyone wishing to understand the complexity of the problem.

Abrahams, H. M. (1959). *Britain's National Parks.* Country Life, London.

Aickman, R. (1955). *The Story of our Inland Waterways.* Pitman, London.

Bracey, H. E. (1963). *Industry and the Countryside.* Acton Society Trust and Faber, London.

Brett, L. (1965). *Landscape in Distress.* Architectural Press, London.

Caborn, J. (1965). *Shelterbreaks and Windbreaks.* Faber, London.

Carson, R. (1963). *Silent Spring.* Hamish Hamilton, London.

Christian, B. (1966). *To-morrow's Countryside.* Murray, London.

Coleman-Cooke, J. (1965). *The Harvest that Kills.* Odhams, London.

Colvin, B. (1947). *Land and Landscape.* Murray, London.

Condry, W. M. (1966). *The Snowdonia National Park.* Collins, London.

Crowe, S. (1958). *The Landscape of Power.* Architectural Press, London.

—— (1960). *The Landscape of Roads.* Architectural Press, London.

—— (1963). *To-morrow's Landscape.* Architectural Press, London.

Darling, F. F. (1947). *Natural History of the Highlands and Islands.* Collins, London.

Edson, E. F. The Countryside in 1970, 1st Conference, 1963 (1964). *Proceedings.* H.M.S.O. (London).

—— The Countryside in 1970, 2nd Conference, 1965. *Proceedings* and *Reports of the Study Groups.* Royal Society of Arts.

Ellis, E. A. (1965). *The Broads.* Collins, London.

Elton, C. (1958). *Ecology of Invasions by Animals and Plants.* Methuen, London.

Ernle, Lord (1927). *English Farming, Past and Present.* Longmans, London.

Fisher, J. (1966). *The Shell Bird Book.* Ebury Press and Michael Joseph, London.

Fitter, R. S. R. (1945). *London's Natural History.* Collins, London.

Fleure, H. J. (1951). *A Natural History of Man in Britain.* Collins, London.

Ford, E. B. (1945). *Butterflies.* Collins, London.

—— (1955). *Moths.* Collins, London.

Forestry Commission. Annual *Reports.* H.M.S.O., London.

Fox, C. (1943). *The Personality of Britain.* National Museum of Wales, Cardiff.

Game Research Association. Annual *Reports.* Fordingbridge, Hampshire.

Gladstone, H. S. (1930). *Shooting Bags and Shooting Records.* Witherby, London.

Goodman, G. T., Edwards, R. W., and Lambert, J. M. (1965). *Ecology and the Industrial Society.* Blackwell, Oxford.

Habitat. Monthly News Letter of the Council for Nature, London.

Herber, L. (1962). *Our Synthetic Environment.* Cape, London.

Hynes, H. B. N. (1960). *The Biology of Polluted Waters.* University Press, Liverpool.

Jenks, J. (1950). *From the Ground Up.* Hollis and Carter, London.

Jones, F. G. W., and Jones, M. (1964). *Pests of Field Crops.* Arnold, London.

Jones, J. R. E. (1964). *Fish and River Pollution.* Butterworth, London.

Laverton, S. (1962). *The Profitable Use of Farm Chemicals.* Oxford University Press, London.

Lockie, J. D., and Ratcliffe, D. A. (1964). 'Insecticides and Scottish Golden Eagles'. *British Birds,* 57, 89–102.

Macan, T. T., and Worthington, E. B. (1951). *Life in Lakes and Rivers.* Collins, London.

Macan, T. T. (1963). *Freshwater Ecology.* Longmans, London.

Martin, H. (1964). *The Scientific Principles of Crop Protection.* Arnold, London.

Massee, A. M. (1954). *The Pests of Fruits and Hops.* Crosby Lockwood, London.

Matheson, C. (1931). *The Brown Rat and the Black Rat in Wales.* National Museum of Wales, Cardiff.

—— (1932). *Changes in the Fauna of Wales within Historic Times.* National Museum of Wales, Cardiff.

Mellanby, K. (1967). *Pesticides and Pollution.* Collins, London.

Middleton, A. D. (1931). *The Grey Squirrel.* Sidgwick and Jackson, London.

Ministry of Agriculture, Fisheries and Food:

(1951). *Toxic Chemicals in Agriculture.* H.M.S.O., London.

(1953). *Toxic Chemicals in Agriculture (Residues in Food).* H.M.S.O., London.

(1955). *Toxic Chemicals in Agriculture (Risks to Wild Life).* H.M.S.O., London.

(1959). *Land Drainage in England and Wales.* H.M.S.O, London.

(1961). *Toxic Chemicals in Agriculture and Food Storage.* H.M.S.O., London.

(1964). *Review of the Persistent Organochlorine Pesticides.* H.M.S.O., London.

(1964). *Review of the Persistent Organochlorine Pesticides (Supplementary Report).* H.M.S.O., London.

(1964). *Report of the Research Committee on Toxic Chemicals.* Agricultural Research Council. H.M.S.O., London.

(1965). *Supplementary Report of the Research Committee on Toxic Chemicals.* Agricultural Research Council. H.M.S.O., London.

Ministry of Housing and Local Government:

(1962). *Water Conservation: England and Wales.* H.M.S.O.. London.

(1963). *Conference on Water Resources in the North-West.* H.M.S.O., London.

(1964). *The South-East Study.* H.M.S.O., London.

Moore, N. W. (1965). 'Pesticides and Birds'. *Bird Study,* 12, 222–52.

Murton, R. K. (1965). *The Wood-Pigeon.* Collins, London.

National Parks Commission. Annual *Reports.* H.M.S.O., London.

National Trust. Annual *Reports.* London.

Nature Conservancy. Annual *Reports,* H.M.S.O., London.
—— (1965). *Report on Broadland.* H.M.S.O., London.
Nicholson, E. M. (1951). *Birds and Men.* Collins, London.
—— (1957). *Britain's Nature Reserves.* Country Life, London.
Pearsall, W. H. (1950). *Mountains and Moorlands.* Collins, London.
Prior, R. (1965). *Living with Deer.* Deutsch, London.
Royal Society for the Protection of Birds and British Trust for Ornithology (1960 *et seq.*).
—— *Report on Toxic Chemicals.* Sandy, Bedfordshire.
Rudd, R. L. (1965). *Pesticides and the Living Landscape.* Faber, London.
Salisbury, E. (1961). *Weeds and Aliens.* Collins, London.
Shorten, M. (1954). *Squirrels.* Collins, London.
Stamp, L. D. (1955). *Man and the Land.* Collins, London.
Stamp, L. D., and Hoskins, W. G. (1963). *The Common Lands of England and Wales.* Collins, London.
Stapledon, G. (1935). *The Land Now and To-morrow.* Faber, London.
Steers, J. A. (1953). *The Sea Coast.* Collins, London.
Summers-Smith, D. (1964). *The House Sparrow.* Collins, London.
Swan, L. A. (1964). *Beneficial Insects.* Harper and Row, New York and London.
Tansley, A. G. (1945). *Our Heritage of Wild Nature.* Cambridge University Press, London.
—— (1949). *The British Islands and their Vegetation.* C.U.P., London.
Thompson, H. V., and Worden, A. N. (1956). *The Rabbit.* Collins, London.
Trist, P. J. O. (1947). *Land Reclamation.* Faber, London.
Vesey-FitzGerald, B. (1946). *British Game.* Collins, London.
—— (1965). *Town Fox, Country Fox.* Deutsch, London.
Waterways Board. (1964). *The Future of the Waterways.* H.M.S.O., London.
Wibberley, G. P. (1959). *Agricultural and Urban Growth.* Michael Joseph, London.
Woolridge, S. W., and Goldring, F. (1953). *The Weald.* Collins, London.